ELEMENTS OF ETHOLOGY
A textbook for agricultural and
veterinary students

D.G.M. WOOD-GUSH
The Edinburgh School of Agriculture,
University of Edinburgh

LONDON   NEW YORK
CHAPMAN AND HALL

First published 1983 by
Chapman and Hall Ltd
11 New Fetter Lane, London EC4P 4EE
Published in the USA by
Chapman and Hall
733 Third Avenue, New York NY 10017
©*1983 D.G.M. Wood-Gush*

Phototypeset by Sunrise Setting, Torquay, Devon
Printed in Great Britain by J.W. Arrowsmith Ltd, Bristol

British Library Cataloguing in Publication Data
Wood-Gush, D.G.M.
   Elements of ethology
   1. Animals, Habits and behaviour of
   I. Title
   591.51   QL751
   ISBN 0-412-23160-3
   ISBN 0-412-23170-0 Pbk

Library of Congress Cataloging in Publication Data
Wood-Gush, D. G. M. (David Grainger Marcus)
   Elements of ethology.
   Bibliography: p.
   Includes index.
   1. Domestic animals—Behavior. 2. Animal behavior.
I. Title.
SF756.7.W66 1983   636′.001′59151   82-23446
ISBN 0-412-23160-3
ISBN 0-412-23170-0 (pbk.)

ELEMENTS OF ETHOLOGY

CONTENTS

experience and the development of sexual behaviour. Social factors affecting sexual behaviour. Some conclusions in relation to farm animals. Points for discussion. Further reading.

PREFACE

The object of this book is to introduce ethology to agricultural and veterinary students. Today ethology covers many approaches to the study of animal behaviour which are connected by one unifying concept: all behaviour must be considered in relation to the ecology and evolutionary history of the species under investigation. This may seem to some to put domesticated animals beyond the scope of classical ethology but, while domestication has involved some behavioural changes, we shall see that much of the behaviour of our species of farm livestock differs little from that of their putative ancestors.

It is assumed that students using this book will already have studied some physiology. It is also assumed that they are, essentially, practically minded and with this factor in mind I have discussed behaviour in terms of its function, introducing the principles of ethology within functional categories of behaviour. In order to best illustrate these principles I have taken examples from a variety of species and not confined myself to farm livestock and domestic animals, for fundamental ethological research with these species has been patchy. However at the end of each chapter I have given a list of papers pertaining to farm livestock so that the principles of ethology can be seen in a more practical context and to develop this approach further I have also added some practical problems for discussion at the end of each chapter.

ACKNOWLEDGEMENTS

I would like to thank Katherine Carson, Alistair and Candace Lawrence, Candace Miller, Ruth Hutton and Peter Wright for their help and comments on the preparation of the typescript. Thanks also to Joy Muir, for helping with some of the diagrams.

1
LEARNING

Before proceeding to discuss the various aspects of animal behaviour it is necessary to consider learning in some detail. Not only is it involved in the perfection and shaping of most behaviour patterns but it is also used as an experimental tool in the study of behaviour. If one wants to know whether an animal can discriminate between two colours, or two tones, or even between two individuals, certain types of learning experiments are involved and a knowledge of how this is done is needed.

Learning has been defined as that process which manifests itself by adaptive change in individual behaviour as the result of experience (Thorpe, 1963). It is conventional to classify learning according to how the learning takes place, but this does not mean that the different categories differ physiologically from one another, although they may do. Our knowledge in this respect is yet meagre.

Learning has been classified into several categories, some of which are considered below.

CLASSICAL CONDITIONING

Conditioning has been defined as the formation or strengthening of an association between a conditioned stimulus and a response, through the repeated presentation of the conditioned stimulus in a certain relationship with an unconditioned stimulus that originally elicits the response. This type of learning became widely known through the work of the famous Russian physiologist, Pavlov. In his experiments, Pavlov used dogs as his subjects. The animal was placed in a restraining harness (Fig. 1.1) and was then presented with meat powder (the unconditioned stimulus, S_1) and almost simultaneously

1

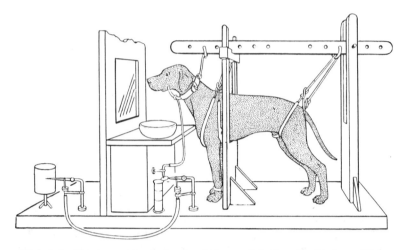

Fig. 1.1 Apparatus used in classical conditioning by Pavlov. (From Hilgard, 1957, after Yerkes and Morgulis, 1909).

presented with another stimulus such as the sound of a metronome (the conditioned stimulus, S_2). Presentation of the unconditioned stimulus, S_1, causes the dog to salivate and after repeated presentations of both stimuli, it was found that the sound of the metronome without the presentation of the powdered meat produced salivation. Diagrammatically the process is shown in Fig. 1.2. For conditioning to occur it is essential that the conditioned stimulus, S_2, should not follow S_1. Also, if the unconditioned stimulus, S_1, starts before S_2 then it should continue after S_2 has stopped. The conditioned reflex is formed by the association of the new stimulus with a reward, which in the case cited was meat powder; the process of reward being called 'positive reinforcement.'

The process of conditioning can be quantified and the following measurements taken:

1. The amplitude of the conditioned reflex, e.g. the number of drops of saliva.
2. The latency of the response. (The time from the presentation of the conditioned stimulus to the appearance of the conditioned response.)
3. The number of reinforcements needed before the first measurable conditioned response appears.

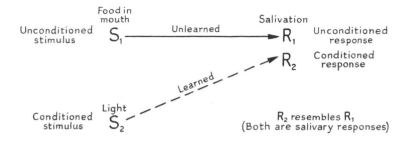

Fig. 1.2 The diagrammatic representation of the relationship between the unconditioned stimulus and the unconditioned response, which does not have to be learnt. The relationship between the conditioned stimulus and the conditioned response is learnt through the pairing of the conditioned and the unconditioned stimuli. (From Hilgard, 1957.)

4. The percentage of trials in which the conditioned response occurs, for even after many reinforced presentations the conditioned response does not occur at every presentation of the conditioned stimulus.

With respect to the last measurement, care has to be taken to avoid confusing the non-appearance of the conditioned response with the phenomenon of extinction, which occurs when the conditioned stimulus is presented repeatedly without reinforcement. Studies have shown that the decrease in response during extinction is not due to a mere passive disappearance of the response, but due to an active inhibition which may spread to other responses not included in the original conditioning. However, extinction is not permanent, for after a while spontaneous recovery may occur.

When a conditioned response to a particular conditioned stimulus has been acquired, other similar stimuli will evoke the response and this is known as generalization. Inhibition also generalizes so that if extinction of the response occurs to one stimulus it will also occur to other similar stimuli. Discrimination between similar stimuli may be brought about by selectively reinforcing only one stimulus amongst trials with other similar stimuli. If the animal can distinguish between the stimuli it will eventually not give the conditioned response in the presence of the non-reinforced stimuli. Discrimination is a very powerful tool for understanding the perceptual world of animals.

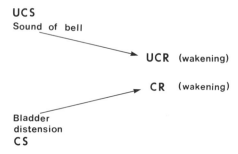

Fig. 1.3   The treatment of enuresis through classical conditioning.

Classical conditioning has been used in the cure of a number of behavioural disorders among humans, including enuresis (bedwetting) (Freyman, 1963). It appears that the fundamental learning deficit in children suffering this condition is the failure to learn to wake up when the bladder is distended, and the treatment aims to form a conditioned response (waking) to the conditioned stimulus (bladder distension). A special electric blanket, wired to an electric bell, is placed in the bed and the system is activated by urination, causing the bell to ring. Since the conditioned stimulus (bladder distension) exists shortly before urination and the ringing of the bell, a conditioned response (waking) is learnt (Fig. 1.3). Classical conditioning has also been found to be useful in the treatment of phobias. In Russia it has been widely used for visceral conditioning in animals and in this respect can be a useful veterinary tool and the reader interested in this aspect is advised to read the article cited at the end of this chapter (Razran, 1958).

OPERANT OR INSTRUMENTAL CONDITIONING

This is the strengthening of an active (operant) response by presenting a reinforcing stimulus if, and only if, the response occurs. This is the technique unconsciously used by people in teaching animals tricks: they reinforce the response they want with a reward, and very soon the animal will give the response at the presentation of the appropriate stimulus. The process is shown diagrammatically in Fig. 1.4.

In general, as mentioned above, the animal is put into a situation in which it has to make a particular response in order to get a reward. For

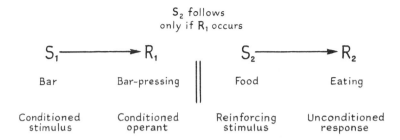

Fig. 1.4 A diagram of operant conditioning. Learning depends upon the strengthening of the response R_1 by reinforcing the stimulus S_2. (From Hilgard, 1957.)

example, the animal is made hungry and placed in a Skinner box (see page 6) which has an automated food delivery mechanism that can be operated by the animal making a certain response. Once it has made that response it is immediately rewarded with some food. Soon the animal comes to make the correct response with more and more assurance. In another context the animal, say a dog, might be rewarded for raising its paw on a command. The closeness in time between the performance of the response and the reward (known as contiguity) is very important, for a lapse of several seconds may impede learning in most cases. The exception appears to be in the learning involved in acquiring language and in avoiding the ingestion of harmful substances. These cases will be considered more fully in Chapter 3. Operant conditioning seems to be more 'natural' than classical conditioning and involves what is sometimes called trial and error learning, in which the animal tries a number of responses and eventually accomplishes its objective, the last response being reinforced by the achievement. The millions of farm animals housed in intensive conditions will use this type of learning when first encountering a nipple drinker. A thirsty animal will be drawn to the nipple drinker by the sight of a drop hanging onto the drinker or smell of water, and as it touches the nipple more water will immediately be released. Because of the contiguity between the response and the reward, the response will be reinforced and rapid learning will occur. Finally, it will have been noticed that in operant conditioning under experimental conditions the reinforcement is positive, i.e. the animal is rewarded for doing the right thing. Even when the design includes electric shocks, the correct response is an avoidance response of some sort. However it must be stated that many bad habits in pets are caused by the owners unwittingly reinforcing the behaviour they wish to avoid. For example a dog yaps when shut out. The owners at first do not want to let it in but, eventually, worn down by the yapping let it in while it is going full blast and reinforce the yapping!

The Skinner box mentioned above is especially designed to use this type of learning to investigate not only principles of learning such as extinction, generalization and discrimination, but also motivation (Fig. 1.5). In the last case one can measure how motivated an animal is by seeing how hard it will work to rectify a particular need.

Once the animal has learnt the relevant response the rate of reinforcement may be altered so that instead of a food pellet being delivered after every correct operant response, it is delivered after a set number of responses. This is known as a fixed ratio schedule of reinforcement. On the other hand the delivery apparatus can be set to

Fig. 1.5 A pigeon in a Skinner box. Every peck she makes at the disc (the operant response) can be rewarded by food (reinforcement). All pecks and reinforcements are automatically recorded. The food delivery apparatus can be seen below the disc to be pecked.

deliver after a varying number of responses, known as a variable schedule of reinforcement. Indeed, in nature it is unlikely that every 'good' response will be rewarded and the ratio of reinforced responses to unrewarded ones will be variable.

In the Skinner box extinction can be studied by ceasing to reinforce the response (Fig. 1.6). To study generalization and discrimination, a particular cue can be inserted in the box, say a light of a particular

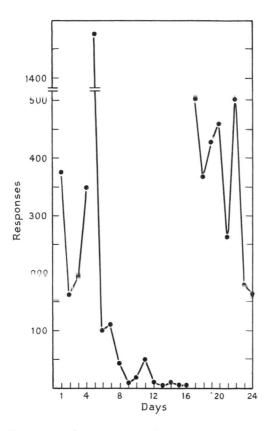

Fig. 1.6 The record of the pecks made by a hen in a Skinner box with food available through pecking at the disc, as well as from an ordinary food trough. On days 1–4 the hen obtained some of her food by pecking at the disc. On days 5–16 only 'free' food was available in the trough, but on day 5 she pecked at the disc at a very high rate, then on days 6–16 the response almost extinguished. (From Duncan and Hughes, 1972.)

wavelength and the response required to get food (pecking at a disc in the case of a chicken or pigeon) is reinforced when the light is on. Generalization can be studied by varying the colours, and discrimination by reinforcing only one colour amongst the presentation of several other colours. The Skinner box is fully automated so that every relevant operant response (e.g. pecking at the disc by a bird or pressing a lever by a mammal) is recorded together with every reinforcement. As we shall see in later chapters this instrument has been widely used to study a large number of specific questions in animal behaviour.

Visceral conditioning has also been demonstrated using operant conditioning. Miller and Banuazizi (1968) treated rats with curare, keeping them alive by artificial respiration. Each time the large intestine relaxed, the medial forebrain bundle was stimulated (rats apparently find this pleasurable, for if allowed to stimulate themselves in this area of the brain they will do so at very high frequencies). This positive reinforcement led to a significant decrease in spontaneous contraction of the large intestine. Similarly rats could be conditioned to contract the large intestine at frequencies well above the base line level. These responses with the large intestine were specific, not involving changes in other organs. When the reward stopped the responses showed extinction. Furthermore, in other experiments heart rate too was raised or lowered significantly and these responses too were very specific.

HABITUATION

Habituation is a simple learning not to respond to repeated stimuli which tend to be without significance in the life of the animal. Responses are omitted. A distance noise causes a startle response but on repetition of the noise the response is no longer given. It must be distinguished from the waning of a response due to muscular fatigue or sensory adaptation, but in many respects resembles the inhibition found in classical conditioning, and is to be distinguished from it by the fact that habituation involves innate responses, or responses presumed to be innate (Thorpe, 1963). In terms of learning theory habituation may be thought of as occurring due to the lack of reinforcement following the response. It is most evident in relation to more generalized and elementary stimuli which make the animal ready for flight. The acquired tolerance of a horse to a saddle and man on its back will involve this type of learning. It will also be referred to again in later chapters in relation to the behaviour of farm livestock.

LATENT LEARNING

Latent learning is learning without patent reward. The acquisition of the knowledge is not displayed at the time of learning, but lies latent. Young children producing a right word in the correct context, not having said it before, display latent learning. In animals it is most apparent in relation to the knowledge they show about the surroundings in which they live. Rats allowed to run a maze without reward, later when tested take a short path to food placed at the far end of the maze. It would be of obvious advantage for an animal in the wild to know its surroundings well before it is attacked by a predator or know where possible food items are when the usual sources run out. Latent learning can thus be surmised to be closely related to curiosity and exploration.

INSIGHT LEARNING

By insight learning is meant the solving of a problem through perceiving the relationships essential to solution (Hilgard, 1957). There is no obvious stage of trial and error, the solution to the problem appears almost instantly. The classical example in animal behaviour is that of a male chimpanzee which was in a cage with a bunch of bananas hanging out of reach. He had a short stick but still could not reach the fruit with it. Outside his cage, out of his reach, was a longer stick. Having failed to get the fruit with the short stick he then gazed about him for a while, then suddenly taking the shorter stick raked the longer stick towards him and, having grasped it, used it to get the fruit (cited by Hilgard, *loc. cit.*).

IMPRINTING

Imprinting became well known from the classical studies of Lorenz (e.g. 1966). In species of precocial birds in which the young are hatched in an advanced state, capable of following the parent, he reported that the young exposed first to a moving object other than the normal parent would follow that object and form an irreversible attachment to it; the young would become imprinted on that object.

Lorenz worked with Grey Lag geese and later work showed that this finding was applicable to duck, chickens and other species of birds. From the early work it was considered that imprinting was a unique form of learning with several unique characteristics: (a) It was limited to a short critical, or sensitive, period. (b) It was irreversible in that once the bond was formed it was stable. (c) It involved learning the broad characteristics of the species rather than the particular characteristics of the parent, so that later the animal would direct its sexual behaviour towards an animal similar to the parent. (d) It was completed long before the responses (e.g. sexual behaviour) to which the imprinted pattern would become linked were established (Thorpe, *loc. cit.*). Following Lorenz's writings numerous experiments were carried out with the young of avian species. The form that these experiments often took was to expose a chick to a moving object as shown in Fig. 1.7 and allow it to follow the object.

Later the chick would be tested for imprinting by exposing it to another moving object to see whether it showed discrimination in favour of the one to which it had been initially exposed and allowed to follow. If it did discriminate then it was considered to be imprinted on the original model. Klopfer (1959) found that in wood-duck (a species that nests in holes), the young become imprinted on the sound of the parents' vocalizations: they showed a form of auditory imprinting.

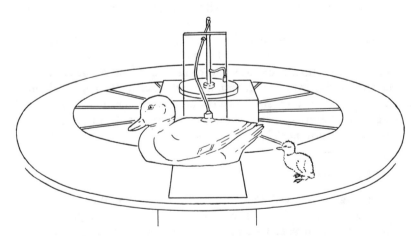

Fig. 1.7   The type of arena used in many imprinting experiments showing the model which can be rotated round the arena allowing the duckling to follow. (From Hilgard, 1957.)

Later studies showed that many of these early views of Lorenz and others were oversimplifications. For example, Guiton (1959, 1961) working with chickens showed that, if a chick was isolated, the sensitive period could be extended and that some male chicks, imprinted onto a cardboard box of a certain colour, when adult directed their courtship behaviour to the cardboard box to which they had been imprinted, and also courted hens of their own breed. Today many ethologists do not consider imprinting to be a unique type of learning, but to be one important aspect of the development of behaviour. Also, filial imprinting, that is the imprinting of the young to a parent figure, has been found to occur at a different time from sexual imprinting which we shall discuss in Chapter 5.

Imprinting-like processes occur in which learning is limited to discrete periods, often early in the life of the animal, and which have very long lasting effects (Immelmann, 1980). These include locality imprinting in which young migratory fish and birds become imprinted to their natal stream or area. Food imprinting is found in insects that are monophagous and whose larvae live on only one species. In parasitic birds host imprinting may occur and Immelmann describes the process in the African wydah bird which lays its eggs in the nests of grass finches. Each species of wydah bird parasitizes only one species of grass finch and the young wydah chicks show an apparently innate similarity in appearance and behaviour to the young of their host species and through an imprinting process are able to parasitize the nests of the correct species. The maternal bond in the female of some mammalian species appears to conform to imprinting in that it is limited to a sensitive period and is irreversible, but it will be considered when parental behaviour is discussed (Chapter 6).

Thorpe (1958) described song learning in male chaffinches which have a song consisting of three phrases. Males brought up in acoustic isolation from a few days after hatching, however do not have the normal song (Fig. 1.8). The first two phrases are often inseparable and the third phrase is usually abnormal. On the other hand young males that, as nestlings, have heard their father or other normally reared chaffinches sing, show the full song, indicating that development of the song is learnt in a limited period early in life.

The examples of imprinting considered here have mainly involved birds but examples of filial imprinting and sexual imprinting in farm livestock have been described by Sambraus and Sambraus (1975). In general though, filial and sexual imprinting in mammals is likely to

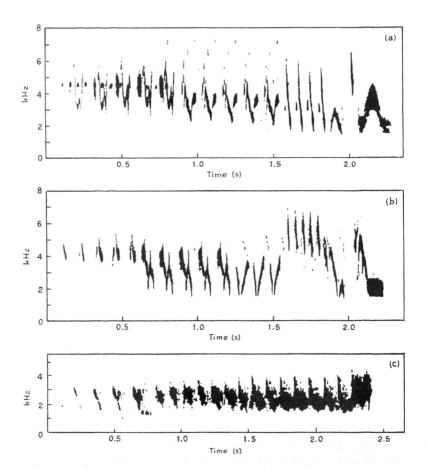

Fig. 1.8 Sound spectrographs of chaffinch songs showing learnt elaboration superimposed on basic innate pattern: (a) and (b) are normal full songs of birds caught in the wild and kept in aviaries, (c) is the basic innate song of a hand-reared chaffinch that had been visually and aurally isolated from other chaffinches. (Thorpe, 1963.)

involve olfactory factors and hence to be difficult to manipulate experimentally. From the work of Scott and his colleagues on the development of social behaviour of the dog (Scott, 1962), social experience, even of a very short duration during the period of 3–10 weeks of age, is necessary for fairly normal social behaviour to occur. However, it must be borne in mind that changes that occur in a particular stage of the development of an animal may be due to maturation rather than to learning. Although imprinting may not be a unique type of learning, as a process it is extremely important in the development of behaviour, and as we have seen, imprinting-like processes may involve many aspects of behaviour. The implications for agriculture are very big, involving mating behaviour, maternal behaviour and possibly even feeding behaviour, and we shall consider them in later chapters.

CONSTRAINTS ON LEARNING

It is very difficult to make a strict comparison of the learning abilities of different species, for their evolutionary histories tend to make them respond differently to a given test. For example the ecology of the pigeon makes it an avid pecker, whereas the chicken, having evolved from a jungle habitat, is a pecker and scratcher and, if required to make a large number of pecks in succession, is unsuccessful because the scratching behaviour interrupts the behaviour. Likewise, one species may rely heavily on visual perception while another may rely more heavily on olfactory information. Any comparison therefore must be extremely cautious.

POINTS FOR DISCUSSION

1. What type(s) of learning are involved when a dairy cow learns to let down her milk to a milking machine?
2. Discuss the process of imprinting in relation to the husbandry of one type of farm livestock.
3. What is the best way of teaching a dog to obey commands?
4. Cite an example in one species of farm livestock in a modern system of production of (a) classical conditioning, (b) operant conditioning, and (c) habituation.

FURTHER READING

Baldwin, B.A. and Meese, G.B. (1977), Sensory reinforcement and illumination preference in the domestic pig, *Anim. Behav.*, **25**, 497–507.

Gustavson, C.R., Kelley, D.J., Sweeney, M. and Garcia, J. (1976), Prey-lithium aversions I: Coyotes and Wolves, *Behav. Biol.*, **17**, 61–72.

Moore, C.L., Whittlestone, W.G., Mullord, M., Priest, P.N., Kilgour, R. and Albright, J.L. (1975), Behaviour responses of dairy cows trained to activate a feeding device, *J. Dairy Sci.*, **58**, 1531–1535.

Lidell, H.S. (1954), Conditioning and emotions, *Sci. Am.*, **190**, 48–57.

Razran, G. (1958), Soviet Psychology and Psychophysiology, *Science*, **128**, 1187–1194.

Warren, J.M., Brookshire, K.H., Ball, G.G. and Reynolds, D.V. (1960), Reversal Learning in White Leghorn Chicks, *J. Comp. Physiol. Psychol.*, **53**, 371–75.

Kratzer, D.D. (1971), Learning in farm animals, *J. Anim. Sci.*, **32**, 1268–1273.

Mugford, R. (1981), The social skill of dogs as an indicator of animal awareness, in *Proc. workshop on self-awareness in domesticated animals* (Wood-Gush, D.G.M., Dawkins, M.S. and Ewbank, R. eds). UFAW, London.

2
SOCIAL BEHAVIOUR

INTRODUCTION

Our species of farm livestock have complex social behaviour. Unlike some other behaviour the basic physiological controlling mechanisms of social behaviour are as yet unknown. Nevertheless it appears to be a motivational system in its own right, not just a by-product of other motivational systems that may bring animals together. However, in social species animals often show group synchronization of behaviour with all the members of a group tending to do the same thing at the same time. Such synchrony is due to the tendency of members of a social species to follow one another, and to the phenomenon of allelomimicry which is a type of behavioural contagion resembling imitation, but not involving the element of conscious copying found in imitation. An example of allelomimetic behaviour in humans is yawning which may quickly pass through a group of people. Social facilitation is defined as any increment or decrease in the measurement of an activity resulting from the presence of another conspecific doing the same thing. Synchronous behaviour may be strengthened by activities having their own diurnal rhythms so that a group may feel hungry or sleepy at the same time.

SOCIAL STRUCTURES

Many animals live in aggregations or in societies. An aggregation lacks the cohesion of a society, and may be described as a group of individuals of the same species comprised of more than just one mated pair or a family, gathered in the same place but not organized or engaged in cooperative behaviour. A society, on the other hand, is a group of individuals belonging to the same species and organized in a cooperative manner. An essential criterion of a society is reciprocal

15

communication of a cooperative nature transcending mere sexual activity.

A society can be described in terms of the following characteristics which measure its sociality, i.e. its properties and processes (Wilson, 1975):

1. The size of the groups composing it; a group being defined as any set of organisms belonging to the same species that remain together to a distinctly greater degree than with other conspecifics.
2. The distribution of the various demographic classes within the population and their inter-relationships, e.g. number of defenders in proportion to the number of reproductively active females.
3. Cohesiveness: this refers to the physical closeness of group members to one another.
4. The amount and pattern of connectedness, by which is meant the number of individuals that can be reached by the signal of an individual.
5. Permeability: the degree to which immigrants are assimilated.
6. Compartmentalization: This is the extent to which the subgroups of a society operate as discrete units and as such it is a measure of the complexity of the society. For example, on being attacked wildebeest flee, each female defending her own young; while in zebras, family units go together and the stallion defends his harem.
7. Differentiation of roles: this is a hallmark of advance in social evolution.
8. Integration of behaviour which measures the degree to which different specialists cooperate to perform a task by integrating their skills.
9. Information Flow: the magnitude of a communication system can be measured in three ways:

   (a) The total number of signals.
   (b) The amount of information in 'bits' per signal.
   (c) The rate of information flow in 'bits' per second per individual and 'bits' per second for the entire society.
10. The fraction of time devoted by the individuals to social behaviour.

Social species may be divided into (a) *contact species*, which are ones in which the animals maintain contact, particularly during resting activities and (b) *distance species*, whose members always keep a certain distance between themselves. A breakdown in this distance, known as 'individual distance', leads to aggression. A particular type of spacing entails the formation of territories which are areas that are defended against intruders; ownership is often advertised vocally or by marking. In some species this territory provides space for breeding and also enough food for the breeding unit. In others it may be used only for breeding purposes. Many species are organized in home ranges which are undefended areas in which they live. This is the system found in hill sheep (Hunter and Milner, 1963).

Many species living in cohesive societies have a dominance hierarchy within their group. Sometimes this may take the form of a linear hierarchy in which an animal is dominant to all those lower in the hierarchy, and subordinate to all those above, or it may take the form, as in wild asses, in which a stallion is dominant to all in his territory, but the other animals appear to be of equal status (Klingel, 1977). Linear hierarchies are very wide spread, having been reported even in invertebrates as well as vertebrates. The dominant animals may contribute most to the gene pool and there is evidence from some wild species, such as the elephant seal, that the dominant male may sire about 80% of the young in one season (Banks, 1977). Animals high in the dominance hierarchy are considered by some to have priority to important resources. However, generalization is dangerous and these points will be discussed in relation to farm livestock later.

COMMUNICATION

For most forms of animal life, communication is vital in most of their activities. For the sake of convenience it may be divided into vocal, olfactory and visual communication, although this division is artificial in that a visual display may, for example, be accompanied by a call.

Vocal communication

In the last thirty years a great deal of work has been done on the vocalization of animals and a number of species previously thought of as silent, such as fish, have been found to vocalize a great deal. The interpretation of the meaning of calls is however difficult. In the

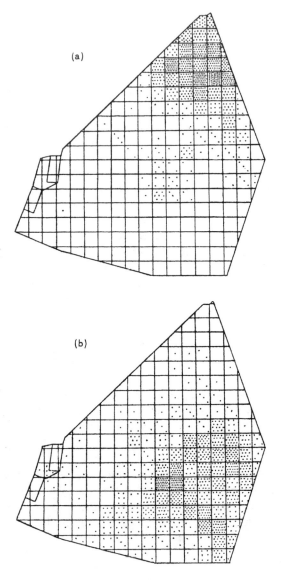

Fig. 2.1 The home ranges of two sub-flocks of Scottish Blackface sheep on a 350 acre hill pasture. (From Hunter, 1964.)

domestic fowl, use of sonographs or audio-spectographs has helped identify twelve different calls for the chick and 19 for the adult birds. However, even identical sonographs may not necessarily mean the same thing, for the context in which the call is made may change their meaning. Nor do we know the communication value of the calls, for very seldom do the scientists working on these calls describe the exact response of other birds.

There is some controversy over the specificity of calls, for some ethologists feel that they are correlated with the general internal state of the animal rather than with specific situations (Kiley, 1972). As we shall see later it does seem that this is true for a number of vocalizations. However there are a number which are very specific. The aerial predator call of the fowl is one of these (Wood-Gush, 1971). Less specific is the crowing of the cock; it seems mainly territorial, for in a group of cocks generally only the dominant ones crow (Salomon *et al.*, 1966).

Even within a species, the calls of subspecies may have no communicative effect whatsoever, e.g. herring gulls from Holland pay no attention to the alarm calls of herring gulls from America (Thorpe, 1961).

An interesting example of communication with an unexpected effect has been reported: Vince (1964) has shown that social facilitation may occur during hatching in some species. In wild birds all the eggs in a nest hatch within a very short period (1–2 h), although some embryos are several days older than the youngest, and while there may be a large variation in pipping, hatching is synchronous. Vince has shown that in certain Gallinaceous birds, such as the Japanese quail, this synchronous hatching is due partly to auditory stimuli, in the form of clicks, from the advanced embryos and she has been able to advance the hatching of single embryos by means of artificial clicks.

Before we can resolve the controversy surrounding the interpretation of calls, we need to adopt a system for identifying the different types of call that may be given. Kiley (1972) classified the vocalization of pigs and cattle on the basis of the following criteria:

1. Length of call.
2. Repetition within a two second interval.
3. Amplitude measured in kcycles or kHz.
4. The pitch of the fundamental (lowest overtone) and change of pitch, represented in cycles per second.

5. Whether the mouth is open or closed (filtration).
6. The tonality of the call. A non-musical call has low tonality and noise level (shown as dark bands of no particular frequency) is high.
7. The presence or absence of fry: This consists of a series of short clicks occurring in quick succession from 0.05 seconds, or less, apart and is found in calls of low tonality. It is probably caused by the lips of the glottis.
8. Whether the call is given on the expiratory or inspiratory phase of respiration.

In the pig 15 calls were thus identified (Fig. 2.2) and their occurrence in a number of different contexts was noted. It was concluded that the 15 calls form a continuum, and none is specific to one situation although some are more characteristic of a particular situation. The level of excitement appears to control the type of call given. Grunts are given when wandering around but never when rushing. On the other hand, the squeal in piglets is given when

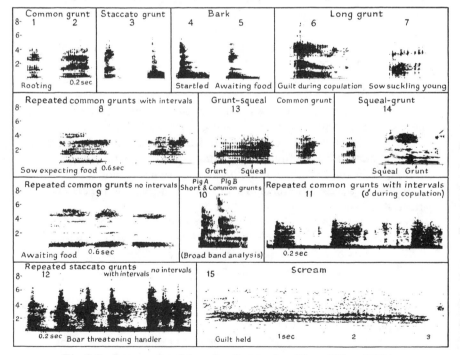

Fig. 2.2   Sonographs of the calls of adult pigs. (From Kiley, 1972.)

rushing or leaping around. The squeals of a piglet can make another piglet leap about, but Kiley considers that the calls of pigs do not, in general, transfer a specific message to other pigs but indicate a change in general motivational state of the calling animal. However the common grunt, given during rooting, may serve to indicate location and would fit in with the forest habitat of the ancestral pigs.

Six different calls were distinguished by Kiley in cattle but, she points out, there are many intermediate stages between the calls which, as in the case of pigs, tend to form a continuum. Similarly they are given in many situations and the call which is given depends on the level of excitement of the animal. As in the case of pigs, increased excitement is accompanied by increased amplitude, frequency, length and or repetition. The calls tend to evoke calling by other cattle, often raising the level of excitement of the receiving animal. Kiley also indicates that, in general, there is a certain amount of variation in length, pitch, tonality and amplitude in the calls of each cow. Although these traits may vary each time a cow gives a call, she can easily be recognized by her calls, for there are far greater differences between the calls of different cows than between the calls of a single cow.

The vocalizations of our common agricultural animals are unlikely to involve much learning, although the animals may learn to vocalize in response to certain stimuli which previously had no effect, or previously had different effects on their calls. Learning may also be involved in the patterning of simple calls as shown in the following example: elephant seals living on islands off the coast of California have a threat call consisting of a number of pulses which are given in bursts. The rate at which these pulses are given vary from one island to another, so that each island has its own 'dialect' (Le Boeuf and Petrinovich, 1974). Recovering from near extinction at the end of the nineteenth century, the species is spreading north and since 1938 has colonized several islands. Extensive tagging of males has allowed Le Boeuf and Petrinovich to examine the possible way in which these 'dialects' may be formed. The evidence, critically sifted, indicates that learning is involved in this relatively simple vocalization so that 'dialects' develop. Young males acquire their characteristic adult pulse rate by entraining and by the time a male reaches puberty it stabilizes. When immature they migrate and probably acquire the pulse rate of the colony into which they immigrate and eventually settle. An island still receiving immigrants would be expected to have a changing dialect and this has indeed been found to be so. However, once an island is fully colonized it has a characteristic pulse rate.

Olfactory communication

Olfactory cues can be very important in communicating information about a sexual partner, a territory, or a dangerous stimulus, as well as advertising information about food sources. To what extent olfactory cues function in the day-to-day integration and synchronization of mammalian societies is not known. However, the following experiment confirms that fear may be transmitted by pheromones (Valenta and Rigby, 1968): Six adult rats were trained to press a bar for food. When they had learnt this, every reward was coupled with air drawn from the cage of an unstressed rat, but every now and again the rat on pressing the bar was given a shock coupled with air from the cage of a stressed rat. Discrimination should therefore lead to a higher probability of bar pressing in 'unstressed' air than 'stressed air'. It was

Fig. 2.3   A pig marking a tree. The movements in marking differ from rubbing and involve the glands used in individual recognition. (Drawn from a photograph by kind permission of Dr A. Stolba.)

found that the presence of 'stressed air' led to an increase in the intervals between bar presses, the mean interval between bar presses in the 'unstressed air' being 1.59 seconds and 8.49 seconds in the 'stressed air' (a highly significant difference). All rats responded similarly and, as some rats were used both as 'stressed' and 'unstressed' air donors, the six were not just responding on the basis of an individual's scent. The involvement of a fear pheromone in the communication system of the back-tailed deer, *Odocoelus hemionus columbianus*, has been reported (Muller-Schwarze, 1974). Apparently it is discharged from the metatarsal glands when the animal is alarmed, causing conspecifics to respond to it when within 20–30 metres of the signaller. Marking is a common form of olfactory communication.

Visual communication

Displays are special postures or facial expressions which convey information to the conspecific beholder and although many are given in combination with vocalizations, (as stated earlier) they will be considered separately here. There are several sorts and in general they tend to display some striking part of the anatomy. Indeed it is considered that during the course of evolution the conspicuous body part has undergone changes to endow it with greater value in a signal.

Threat and appeasement displays

These are two very important classes of displays, not only from the intra-specific viewpoint, but also with respect to good husbandry and our understanding of animals. Threat displays are very common during encounters at the common boundaries of territorial males, and in a social hierarchy a dominant animal often threatens an inferior one by means of a display. Generally it is felt that they serve to reduce actual combat, for when threatened by a superior, the inferior individual within a hierarchy usually gives a submissive display (see Fig. 2.4), or moves away. Many threat postures appear to be derived from conflict between attack and escape and we will discuss this more fully in Chapter 9.

Appeasement displays, like threat displays, act to decrease actual combat. Generally they are of two sorts: those that simulate sexual submission and those that tend to hide any stimuli that are used in

Fig. 2.4   Wild dogs on the Serengeti plains in Tanzania. The bitch on the left is performing the greeting ceremony to the dominant male which has just returned from the hunt. The bitch's display contains several submissive features. (From Wilson, 1975.)

threat, often resembling the behaviour of immature members of the species. Observations in the wild indicate that they are very efficient in decreasing fighting and in saving the inferior animal from mutilation. In intensive agricultural conditions, on the other hand, they are often less successful, and in many groups one may see an animal in a permanent submissive posture. Although the reduction in aggression shown by a dominant animal in response to an appeasement or submissive display is of obvious advantage to the inferior animal, the advantage gained by the dominant animal is not so clear. However it must be recognized that a male may threaten his own mate or his own young from time to time, and if he failed to respond to their appeasement behaviour then he would endanger his own genetic fitness.

Walther (1974) has described two other types of displays in

ungulates: dominance displays and space claim displays. In dominance displays animals do not demonstrate their readiness for immediate attack or defence as in threat displays, nor do these displays refer to weapons. Instead the animals demonstrate their size or striking features such as colour patterns, manes, etc. The most common ones are erected postures and broadside presentations. Space claim displays, as their name implies are used in relation to space but their communicative value is doubtful. The most striking example cited by Walther (*loc. cit.*) is the grazing ritual of the Thomson's gazelle, performed usually by males owning adjacent territories. They graze along the common boundary going through a frontal position (facing one another), a parallel, or reverse parallel position and hind quarters to hind quarters position, grazing all the time and moving along the boundary. The process may last from several minutes to half an hour.

THE DEVELOPMENT OF SOCIAL BEHAVIOUR

The development of social behaviour in our main agricultural species has not been investigated to any great extent except in the fowl. One of the first reactions of a newly-hatched chick is to press against something warm, and it is possible that this stimulus–response interaction may be the foundation of all social behaviour in the fowl, for in an incubator or under natural conditions a chick might often huddle against its siblings or its dam and then will associate them with 'reward' and tend to follow them. Newly-hatched chicks and ducklings may also respond to a 'sense of enclosure' as well as warmth. In addition to this attachment to an object that supplies warmth and/ or a sense of enclosure, it is known that certain stimuli will evoke approach in chicks. For example, chicks that have had their ears plugged from hatching will, when these plugs are removed, go in the direction of a clucking hen which is hidden (Spalding, 1873). Indeed, young chicks will also approach a tapping pencil if it is tapped at the same frequency as the tidbitting calls of a broody hen.

From the initial approach the process of imprinting ensues, with the effects that have already been discussed. Chicks kept socially as in a commercial brooder will probably imprint on one another and will show escape reactions to strange objects from the third day after hatching. Furthermore, they will discriminate against strange chicks from about 10 days of age (Guhl, 1964). Allelomimetic behaviour and social facilitation appear at about one week of age. A form of play behaviour called frolicking, in which a chick suddenly rushes across a brooder and stops without performing any specific behaviour pattern, is affected by allelomimicry at this age and by 2 weeks very marked allelomimetic behaviour and social facilitation are found. For example, in one experiment chicks having learnt to run down an alley to a social lure were found to run significantly faster when run in pairs than when alone (Smith, 1957).

Aggressive behaviour appears in the chick during the second week but the social hierarchy or peck-order is only formed at 7–8 weeks in male groups and at 9 weeks in female groups. Basically the peck-order is a linear hierarchy, but often 'triangular' or 'rectangular' relationships may occur.

$$
\begin{array}{ccc}
 & A & \\
\nearrow & & \searrow \\
C & \leftarrow & B
\end{array}
\qquad\qquad
\begin{array}{ccc}
A & \rightarrow & B \\
\uparrow & & \downarrow \\
D & \leftarrow & C
\end{array}
$$

Implicit in the idea of a peck-order is the notion that birds are able to recognize one another. As imprinting experiments have shown, chicks are able to distinguish between hens from a very early age and they can single out strange chicks from the age of 10 days, although this does not necessarily mean that they can recognize individuals in their own group. However it seems unlikely that the delay in forming a peck-order is due to their inability to recognize individuals. More likely it is due to the fact that their sparring encounters do not lead to clear-cut results until they are able to deliver stronger pecks at one another.

Turkey poults are also easily imprinted and show sexual imprinting and have a social hierarchy similar to that of fowls. Sparring occurs only sporadically in young turkeys up to the age of 3 months and then there is an increase in fighting and, finally, a hierarchy is formed at 5 months. Each sex forms its own peck-order as in fowls, with the males generally dominant to the females.

In mammals the mother–young relationship is extremely important, although under laboratory conditions young animals will develop adequate social behaviour if deprived of the mother but given the company of similarly aged conspecifics (Maple, 1975) and this probably also applies in farm conditions. However, apart from nutrition, the mother provides security and comfort. In an experiment with Rhesus monkey infants reared with surrogate mothers, the nature of the infant–mother bond was examined. The hypothesis tested was that the bond depended on the nutrition provided by the mother (Harlow, 1959). Eight newly-born monkeys were each kept in a cage in which there were two wire cylinders, each with a painting of a monkey's face attached to it; one of the cylinders was otherwise covered in towelling (Fig. 2.5). In the case of four of these monkeys the milk bottle was attached to the cylinder without the towelling and in the other four it was attached to the one with the towel covering. The two groups drank the same amount of milk and grew at

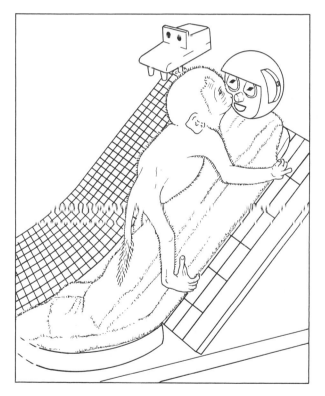

Fig. 2.5 A baby Rhesus monkey with two mother-surrogates. It is clinging to the 'mother' with the soft towelling. This was done by all the infant monkeys even when their milk came from the wire 'mother'. (From Harlow, 1959.)

the same rate. However the infants spent much more time clinging to the towel 'mothers' than to the wire 'mothers'. Those that got their milk from the wire 'mothers' only went to them to drink and never for other purposes. The bond therefore does not appear to be dependent on nutritional reward. When a strange object was added to the cage in order to give the young monkey an emotional stress, the monkeys all sought the towel mother to cling to. If put into a strange cage the young monkey just huddled in a corner and the presence of the wire mother did not change this behaviour. On the other hand, if the towel mother was present, the monkey, after initially clinging to it, began gingerly to explore the cage – first with a hand touching the towel mother and then with a foot touching her. It seems therefore that in young primates maternal contact is very important. A similar relation appears to exist in pigs, for on disturbance young piglets with the sow will run to her and start a sequence of suckling behaviour. Teat-seeking behaviour in newly born lambs is not abolished by the delivery of 600 cm^3 of colostrum by tube into the stomach (Alexander and Williams, 1966). Whether this sucking, like that of the piglets, is due to anxiety is uncertain and other explanations cannot be ruled out.

Ungulate species may be divided generally into 'followers' and 'hiders' according to post-natal behaviour of the young (Lent, 1974). The followers immediately follow the mother after an initial period in which she licks and suckles her young; the hiders, on the other hand after being licked and suckled either find a hiding place where they lie hidden and the mother goes off for long periods, or in some species, such as the feral goat, the mother leads the kid to a hiding place. Lent refers to the first type of hiding as 'lying-out'. In our farm livestock, sheep and horses are followers, whereas pigs may be classified as hiders in that they stay in the nest for the first few days. Calves are also hiders in that they do not follow the cow. However, at an early age they tend to be social hiders in that they will tend to congregate and lie close together in a form of creche behaviour. The cows come back to feed them but there is some disagreement as to whether a 'guard' cow is left with the creche. The caring of the young of other individuals (altruistic behaviour) is fairly common in animals and the subject is discussed fully by Wilson (1975) and by Manning (1979).

The fighting of new-born piglets for teats is not their only social behaviour. They have naso-nasal contact with the sow and have frequent interactions with litter-mates unconnected with suckling. Fighting, however, is more frequent in intensive conditions than in semi-natural conditions. Allelomimetic behaviour is shown from an early age; it is commonplace for them to start suckling the sow when other litters are doing so and this is probably allelomimicry. Litters observed under semi-natural conditions spend much time in play, which is considered to be important in the development of social behaviour (Wilson, *loc. cit.*). By two weeks they show most of the adult agonistic behaviour, but no hierarchy is formed from their encounters. In the litters investigated up to the age of 10 weeks in semi-natural conditions, piglets tended to spend most of their time with certain litter-mates, thus showing more subtle relationships than had been reported from intensive rearing conditions (Fig. 2.6 and Hutton *et al.*, 1980).

Individual penning affects the later social behaviour of calves; such calves occupy the lower ranks in the social hierarchy when grouped with socially reared calves. They also tend to spend more time with one another and are less likely to initiate competitive interactions such as butting, pushing and nudging (Broom and Leaver, 1978).

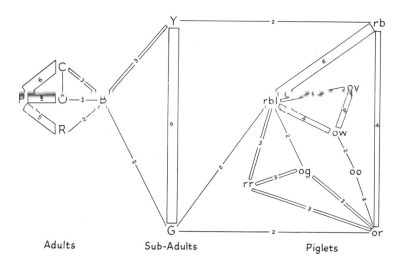

| Adults | Sub-Adults | Piglets |

Fig. 2.6 Sociogram between pigs in an ecologically rich enclosure. B, boar; C, P, O and R, sows; G, sub-adult gilt; Y, sub-adult boar; rbl, rr and rb, unweaned progeny of R; og, ow, oy, oo and or, unweaned progeny of O. The relationships were based on eight different types of social behaviour. (From Hutton, 1979.)

Play

Although most observers feel that they know when an animal is playing, it is very difficult to characterize play in an exact way. Hinde (1970) has listed the following characteristics which, however, are not necessarily found together in any one bout of play:

1. Incomplete sequences of behaviour.
2. There may be the elaboration of new motor patterns specific to the context of play. A piglet, for example, may run and whirl round on its fore-paws.
3. Diverse types of behaviour are intermingled so that predatory and sexual patterns may be mingled.
4. The intensity of the movements may not be in keeping with that of the sequence as a whole. For example the leaping of a kitten in play may be exaggerated.
5. Social play is usually responded to as though not serious by other animals and is often preceded by a special signal. Analysis of data from coyotes (Bekoff, 1976) shows that if a threat signal is preceded by a play signal, it is significantly less likely to be followed by submission than is a threat signal not preceded by a play signal. Communication of this sort which indicates 'what follows is play' is termed metacommunication; it qualifies the subsequent displays.
6. The behaviour patterns employed are elicited by a wide range of stimuli many of which would never be sufficient enough to elicit the behaviour in normal circumstances, e.g. the aggressive behaviour elicited in a kitten by its mother's tail.
7. Amongst higher mammals the play of the young is often elicited by an older animal.

Few of these characteristics have been examined in detail. It is possible that close examination will show that play is not a single system of behaviour (Bateson, 1977). These characteristics apart, play has been divided into various types: aggressive play, manipulative play in which objects are handled, social play and diversive play in which the animal appears to be exploring some aspect of its environment (Fig. 2.7).

Since play is very common in young animals it has been considered to be important for the full development of the social behaviour of animals. Bekoff (1976), for example, while acknowledging that there is little firm evidence, suggests that social experience such as play may

Fig. 2.7 Play in piglets. (Drawn from a photograph by kind permission of Dr A. Stolba.)

be essential in order for the animal to acquire the communicatory skills of its species and it may help an animal to predict how other animals are going to behave and so to alter its behaviour accordingly. The classical method of testing such hypotheses is to deprive animals of play, but this generally involves more than just deprivation of play. Likewise it has been suggested that play may aid prey-killing in predatory species. Vincent and Bekoff (1978) investigated the influence of previous playful and agonistic experience with con-specifics on the later success in killing prey by coyotes. The sample was very small, but they found that success in killing prey was not correlated with the frequency of participation in playful or agonistic interactions, nor, in the context of play, with the frequency of occurrence of motor patterns used in the prey-killing, except for one such motor pattern. Nevertheless the authors suggest that play may improve perception and motor coordination.

Brownlee (1954) suggested that vigorous play functions to stimulate muscle, bone and the cardio-pulmonary system, thus serving as physical training and exercise for muscles used later for reproduction and survival. Fagan (1976) has recently reviewed the phenomenon of play from this viewpoint and pointed out that exercise gives greater physiological benefit in young than adult subjects so that there would be selection for play to occur in early life when it would also not divert the animal from other activities. However, he points out that this function cannot be ascribed to all types of play such as manipulative, diversive or social play.

THE SOCIAL BEHAVIOUR OF SOME AGRICULTURAL SPECIES

Modern methods of intensive husbandry will affect the social behaviour of our farm livestock, but before considering this it is necessary to know more about their full range of social behaviour. In order to do this it is necessary not only to consider their behaviour in the more conventional husbandry systems, in which they were kept for many generations, but also to study the behaviour of feral populations or, preferably, of modern stock kept in semi-natural conditions, in which restriction is at a minimum and the environment is rich enough, ecologically and socially, to provide a wide range of stimuli for social responses. Finally, it is also instructive to consider the social behaviour of the putative ancestral species and often closely related species. From these sorts of studies we will be able to see the derivation of behaviour patterns and functions of those that may puzzle us. Finally it may help us to understand the aetiology of deleterious behaviour patterns.

Fowl

Studies on a feral population of fowls on Northwest Island, off the coast of Queensland (Fig. 2.8) indicate that its social structure is basically territorial (McBride *et al.*, 1969). Each territory is occupied by a dominant cock, 4–6 hens, pullets and several subordinate males and, within such a group, a hierarchy is found. Some of the subordinate males may show varying degrees of territoriality; some defending a very small territory into which the dominant male may intrude. Normally the territories provide the group with all their needs, but during the breeding season the hens with broods become solitary and occupy distinct home-ranges. At this time the group may be reduced to the territorial male and one female. It would seem that space alone is not the factor that leads to territorial behaviour in the fowl, for observations on a small group of about nine adult birds that

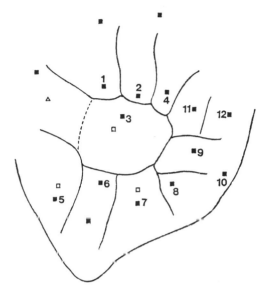

Fig. 2.8 The home range system with roosting points in a population of feral domestic fowl on Northwest Island, Queensland. ■ Roosting points; □ Secondary roosting points; △ Abandoned roosting points; — Normal home range boundaries; ...... Boundaries not observed. (From McBride *et al.*, 1969.)

included several cocks, living in the wild on an island off the coast of Scotland, revealed no territorial behaviour (Wood-Gush *et al.*, 1978). Possibly it depends on both space and a high ratio of females to males, so that harems can be formed and separated.

Invariably a peck order is soon formed when strange birds are put together, the males forming their own and the females theirs, with the males generally being dominant to the females. As stated, the concept of a social hierarchy implies an ability of individuals to recognize one another, i.e. to distinguish and remember individuals. Although it is probably true that the peck-order in small groups of chickens involves individual recognition it is not necessary for all relationships to be founded on it. In large groups one might get the impression of a peck-order but some birds may simply be avoiding birds larger than themselves and threatening smaller ones. In addition a large red comb acts as a strong stimulus for fleeing (Guhl and Ortman, 1953). While

Candland (1969) found that cocks could distinguish other cocks by their combs, other attempts to discover the features by which fowls recognize one another have not yielded clear-cut results (Guhl and Ortman, *loc. cit.*). Features of the head, neck and comb appear to be important (Fig. 2.9).

A number of genetic studies have been carried out on the practical effects of social dominance in laying hens, although the authors have sometimes referred to this trait as aggressiveness. Komai *et al.* (1959) measured the social ranks of birds by counting all fights, pecks,

Fig. 2.9 Alterations made to test for patterns of recognition. Bottom right figure shows hens avoiding their pen-mate after red feathers had been added to her neck. (From Guhl and Ortman, 1953.)

threats and avoidances. Three strains of White Leghorns and one each of Black Australorp, Rhode Island Red and White Plymouth Rock were used. The trait was found to be heritable with mean intra-strain heritability estimates of 0.30 and 0.34. Using the same measures, Tindell and Craig (1959) determined social dominance in different families in small flocks of several breeds, and found that high ranking pullets gained weight significantly faster up to 5 months of age and had significantly better egg production for the first 4 months than lower ranking birds. In another study Craig *et al.* (1965) carried out bidirectional selection of cockerels for five generations for social dominance and dominance scores in paired contests. Large strain differences for the two traits were produced in each of two breeds (White Leghorns and Rhode Island Reds). However, when the females from the selected strains were placed in flocks composed of females from the different strains their peck-order status appeared to be reduced compared to their ranks based on initial paired contests. For example, in one 'dominant' strain the females won 87% of the encounters with females from a 'non-dominant' strain but when these females were housed together the dominant strain dominated only 65% of the non-dominant strain, suggesting that the motivation for this trait can vary according to circumstances. Contrary to earlier findings (Wood-Gush, 1971), Craig *et al.* (1977) found no effect on mating frequency of social status in cockerels.

The postures of the cock in agonistic encounters have been fully described by Wood-Gush (1956). Most of them are of doubtful communicative value and so will not be discussed here. The waltz, first described by Davis and Domm (1943), however does act as a clear-cut threat display in male encounters, for Wood-Gush (*loc. cit.*) found definite responses by the recipient in 79 out of 91 cases analysed. In this display the far wing is dropped and the cock advances sideways with shuffling steps or circles round the other bird with the same gait (Fig. 2.10). Variable intensity is found: the wing may be fully or only slightly lowered and the distance moved may vary. Also the neck may be extended or slightly retracted.

Wing-flapping may be performed by an aggressive cock, or by a retreating one. Wood-Gush (*loc. cit.*) treated all wing-flaps as a single type of posture but this is open to question. In some cases it is very striking: the male stands at his full height and flaps the wings so vigorously that they clap together loudly. This type of wing-flapping probably does have some communicative value as a threat display or,

Fig. 2.10   A Brown Leghorn cock waltzing to a female. Note the lowered outer wing and shortened neck. (From Wood-Gush, 1970.)

at least, as a dominance display. Feeble wing-flapping associated with escape is unlikely to have communicative value.

Submission in the male is generally shown by some degree of crouching and retraction of the neck. Sometimes one may see a bird in a more or less permanent submissive posture, huddled with its neck fully retracted. In the female, submission is a full crouch similar to that adopted in copulation but, in developed peck-orders, threat is generally limited to the intention movement of pecking at the subordinate by the raising of the head to a position for striking and submission is shown by a quick lowering of the head and retraction of the neck.

Nineteen calls have been catalogued for the adult (Wood-Gush, 1971), and the calls of the immature bird have been fully described by Guyomarc'h (1962). However here we shall not be considering calls given in courtship or between the hen and hen chicks. The calls that undoubtedly transmit information are the aerial predator and ground predator calls (Fig. 2.11). Both almost invariably alert the flock, and

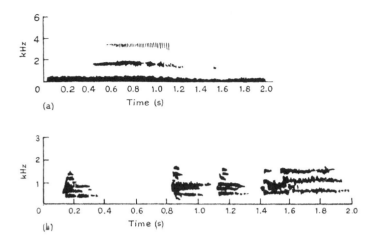

Fig. 2.11 (a) The call given in response to an aerial predator. (From Konishi, 1963.) (b) and (c) The call given in response to ground predators by the domestic fowl. (From Collias and Joos, 1953.)

have been described by Collias and Joos (1953). The aerial predator call which was mentioned earlier is a raucous screech that can be released by throwing a cloth into the air. The ground predator call is initially segmented and then ends on a sustained note of more than half a second long, and is given in response to the approach of a man or animal.

Pigs

Weaners kept in small groups form a social hierarchy in which individual recognition is involved. Ewbank and Meese (1974) investigated the importance of vision and scent in the formation of the hierarchy. When opaque lenses were fitted over the pigs' eyes the pigs formed a stable hierarchy. However when hoods with opaque lenses were fitted over the head, covering the face, hierarchies were not formed, suggesting that scent from the glands on the side of the head are important in recognition. However, hierarchies were formed with hoods in which the eyes were uncovered, indicating that vision can be used in the absence of the cues from the normal scent glands (Fig. 2.12).

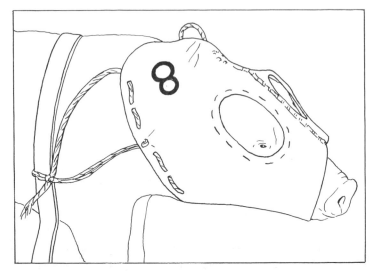

Fig. 2.12 A pig fitted with a hood in which eye-holes have been made and then covered with transparent plastic. With these masks, which interfered with both visual and olfactory cues, individual recognition was impossible for the pigs. (From Ewbank and Meese, 1974.)

The social behaviour of the domestic pig seems to resemble, in all important respects, that of the European wild boar, *Sus scrofa*, when the domestic form is allowed to live in semi-natural conditions (Stolba and Wood-Gush, 1981). The behaviour of the wild boar has been described and compared with that of other closely related species of Suidae by Fradrich (1974). The basic social unit of the wild boar is the mother and her litter. Weaning takes place when the piglets are 3 months old at which time two or more sows assemble and form a family group. In the domestic population studied by Stolba and Wood-Gush, the sows tended to join together when their litters were 2–3 weeks old, but this may have been due to the relative smallness of their habitat compared with the home ranges of the wild boar sows. The wild boar family group then remains unchanged until the mating season when the boar joins them and the young males leave. The boar remains only until mating is complete. Near the time of parturition the young of the previous year are left by the sow who seeks a site and builds a nest. However, the young of the previous year, particularly

Fig. 2.13 Pigs grazing in the Edinburgh University 'Pig Park'. Note the spacing between pigs. Some of the pigs tended to have special neighbours under these conditions. (Drawn from a photograph by kind permission of Dr A. Stolba.)

the females, may then later rejoin the sow and stay with her and her new litter until they are sexually mature, at 8–10 months. A very interesting feature of the home range of the wild boar is characterised by having a number of sites for different activities: nesting sites at which they construct sleeping nests, wallows, feeding sites, rubbing sites at which they rub themselves against certain trees, etc. All these sites within the home range are connected by paths which are regularly used. No area is defended except for the nest. All this behaviour is found in the domestic pig in semi-natural conditions (Stolba and Wood-Gush, 1981). It is particularly interesting to note that the nest built at the nesting sites by the domestic pigs in this study were well sheltered and commanded good views of the approaches. Also, like the wild boar sow, the domestic sow builds an elaborate nest away from the family unit before parturition.

The vocalizations of the domestic pig were discussed earlier and are shown in Table 2.1. Their main displays outside courtship are described below.

Table 2.1 Situations in which the various calls of the pig occur.
++ indicates particularly characteristic of this situation, (+) rare occurrence, blank – not recorded.

Situation	Common grunt	Staccato grunt	Long grunt	Bark	Repeat common grunt & interval	Repeat staccato grunt & interval	Repeat no interval	Grunt squeal	Squeal grunt	Stuttered squeals	Chirrup squeals	Squeal	Scream	Repeat scream
Greeting Confident	+	+	+		+	++	+			+	+			
Greeting of equals	+	+	+		++	++	+			+	+			
Defence threat	+	+			+		+	+	+	+	+	+	+	
Aggressive threat						+	(+)	+	+	+	+	+		
Fear	+	+			+	+	+	+	+	+	+	+	+	+
Close contact retaining	+	+	+		+	+	+							
Tactile stimulation	+	+	++		+	+	+							
Isolation	+	+			+	+	+	+	+	+	+	+	+	+
Startle				+								+		
Pain/fear	+	+		(+)	+	+	+	+	+	+	+	+	+	+
Frustration	+	+	+	(+)	+	+	+	+	+	+	+	+	+	+
Anticipation: pleasant	+	+	+		+	+	+	+	+	+	+	+		
Anticipation: unpleasant	+	+	+		+	+	+	+	+	+	+	+		
Disturbance (unidentified)	+	+		(+)	+	+	+	+	+	+	+	+		

Note: The columns "Repeat common grunt & interval", "Repeat staccato grunt & interval" and "Repeat no interval" fall under the grouping header "Repeat"; the columns "Stuttered squeals", "Chirrup squeals", "Squeal", "Scream" and "Repeat scream" fall under the grouping header "Sub-adult calls".

Threat displays

1. Head lowered. This is given by an animal that is confident. The snout is about 5 centimetres above the ground, the shoulders are high and the back straight. The front legs are fairly stiff and the ears forward.
2. Head up. This is a more aggressive display; the head is higher than horizontal, the mouth closed, the ears erect or back and the approach is sideways. The legs are stretched and straight as if the animal were trying to increase its size. The back is straight.
3. Arched back. This is a still more aggressive display and is identical to 'head up' except that the back is arched.

The defensive threat. When approached by a threatening pig the other may adopt the defensive threat in which the front legs are stretched while the back legs are ready to spring away. The head is low, the shoulders high and the ears generally back. The tail is up and

Fig. 2.14 Adult pigs arranging the communal nest in the late afternoon. When the nest is complete and the adults are settling in the young piglets will join them. (Drawn from a photograph by kind permission of Dr A. Stolba.)

snapping movements are made with the jaws. Also the pig may run straight at its opponent and give it a push or a bite.

Amongst pigs that know one another different approaches are used:

1. Approach I. The legs are stretched, the approach is frontal and the head is swayed horizontally. The back is straight and the tail is often wagged.
2. Greeting. This is friendlier than Approach I. The pigs make naso-nasal contact often standing face to face giving short deep repeated grunts.
3. Grooming. This is the friendliest of interactions. One pig may nose the arm pit area of the other pig or lick its ear or nibble at an ear or side of the face. On being nosed a pig may lie down and expose its belly or side.
4. Submission. There is no real submission posture but the head is turned slowly to the side as an intention movement to flee.

Fig. 2.15   One sow grooming another which she knows extremely well. This is a sign of deep acquaintanceship. (Drawn from a photograph by kind permission of Dr A. Stolba.)

Cattle

The social unit of the cattle of the Carmargue is described by Schloeth (1958) as consisting of a polygamous herd without a fixed territory and generally having several adult males, some of which are castrated. There is a male hierarchy and a female one, with males of one and a half to two years of age being between the two hierarchies. The surplus males impose a strain on the herd and the hierarchy is never definite, continually undergoing changes. The Chillingham cattle, which are a herd dating from at least 1225 AD, have an interesting structure. The females and young stock roam freely over the entire parkland of 150 ha; while the bulls are in three groups with different home ranges. Young bulls may change from one bull-group to another, or to the main herd (Hall, 1979).

Domestic cattle form a hierarchy which can be very stable. Sambraus (1977) reported on such a one that he had observed over an eleven-year period. Rank-order was highly correlated with age. Particular relationships were found in that certain animals tended to lie next to, and to groom, certain other ones (Fig. 2.16). 'Leadership' is unconnected with social dominance but as Leyhausen and Heinemann (1975) point out, as cattle have virtually 360° vision, it is possible that the animals are being 'pushed' from behind by a more dominant animal. They have also suggested that cattle may follow in an attempt to keep the distances between members of a herd more or less constant. Finally, the social attractiveness of an animal may also affect movement by causing following. In herds that have been established for some time, rank-order has been correlated with age (see above) or with body weight, but not necessarily with the possession of horns (Bouissou, 1972). However, in newly formed social groups both body weight and horns are important (Bouissou, *loc. cit.*). A lack of aggression may be found between members of a herd and such preferential relationships are stronger between females that have been reared together from birth than those assembled at 6 months of age or later, suggesting a critical period for the development of these relationships (Bouissou and Andrieu, 1978). No correlation has been found between dominance rank and milk production (Schein and Fohrman, 1955, Dickson *et al.*, 1967).

The vocalizations of cattle were discussed briefly earlier and fuller details are given by Kiley (1972). The displays of cattle have been discussed by Schloeth (1956, 1958), although mostly in very general

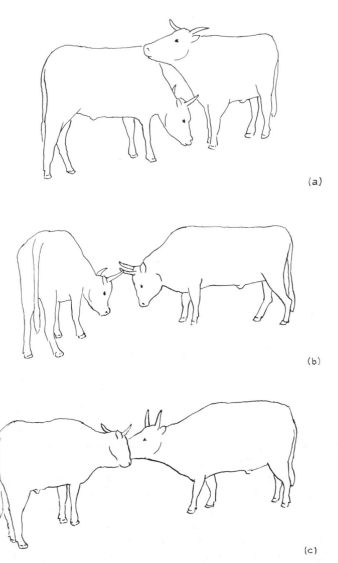

Fig. 2.16 Licking between two castrated bulls. (a) The animal on the right is the dominant one and is licking the shoulder of the other. (b) After a small movement of the head of the animal on the left, the two cross their horns. (c) Agreement. The dominant animal resumes licking the inferior animal on the cheek but this time stands somewhat further away. (From Schloeth, 1956.)

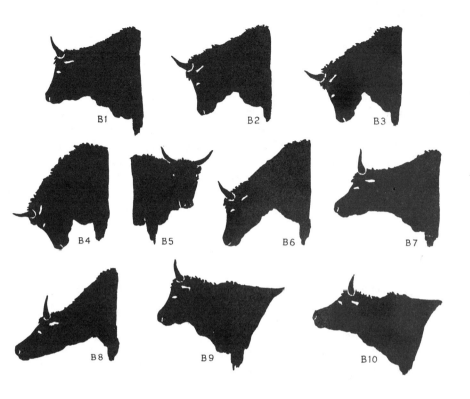

Fig. 1.17 B1: Normal position I. The head in a neutral position. B2: Lateral position no. 1. A milder form of threat. The animal places itself at a right-angle to its adversary. B3: Lateral position no. 2. The form is very pronounced and the animal is close to combat. The animal places itself at a right-angle to its adversary. B4: Combat position. This is a threat or stance for defence. B5: Twisting of the head. A threat before or during a fight; the animal presents its side to its adversary. B6: Normal position no.2, browsing. An expression of social inactivity. B7: Approach position no. 1. A stance of greater assurance showing the intention to enter a group or make social contact (e.g. to play, fight, etc). B8: Approach position no. 2. A stance of less assurance often used by animals of lower social rank. B9: Alert flight position. The line of the back is concave and the base of the tail elevated. This is used by a defeated animal. B10: Position of surveillance. This is a signal for the members of the group to be prepared for flight. The line of the back is concave and the back of the tail raised. (From Schloeth, 1958.)

terms. Facial communication is not found in cattle and the role of the ears is minor. As in pigs the general aspect of the body is important, the relationship of the head to the neck and body being particularly important. Two pivots are involved, the base of the neck and the occipital region which is responsible for the changes in the position of the head. The line of the back, the position of the legs and the tail are also involved. The horns themselves are not important, although in threat postures the head is held so that the horns may be used. On the other hand when the animal is uneasy, as shown in Fig. 2.16(c) the horns are upright, the muzzle forward and the hind legs are forward so that the animal can readily spring away. In the alert posture the head is slightly elevated (see Fig. 2.17, B9 and B10), the line of the back is concave and the base of the tail is raised.

Schloeth (1956) describes several patterns in detail:

Scraping the ground (Fig. 2.18). The head is held low with the muzzle a few centimetres above the earth. After sniffing intensively the animal scrapes the soil with each of its front legs. In a good number of cases, the bull raises the head slightly after some scrapes and turns toward the side of the hoof still scraping. It can be accompanied by the characteristic calls, or by defaecation or, in the case of males, by an erection. It is often followed by 'rubbing the neck'.

Fig. 2.18   Scraping the ground. The animal scrapes with the left leg and with the head slightly inclined to the side of the foot used for scraping, and sniffs the ground. Note that the tail is in the position characteristic for defaecation. This may be performed in positions of threat and aggression. (From Schloeth, 1956.)

Fig. 2.19 Rubbing the neck. This display as in the case of 'scraping the ground' is important with maintaining rank and signifies threat and aggression. (From Schloeth, 1956.)

Rubbing the neck (Fig. 2.19). The animal kneels and rubs its jowls, throat or neck on the ground. Mostly the three parts are rubbed one after the other. It is accompanied generally by a characteristic call, or defaecation, and also sometimes by an erection. As in the previous display the animal is highly excited.

Digging the ground with the horns. This sometimes follows the two previous displays and is often combined with them. The horns are lowered to the ground and the front legs may be bent. It is often, but not always, accompanied by a characteristic call.

The lateral position. The buck is slightly arched, the head slightly lowered and the horns are placed in a position for immediate combat. The animal makes a number of small lateral movements with the posterior part of its body, placing itself at right angles to its opponent and thus displays its flank.

The threat displays are intention movements to fight and submissive postures are derived from the same context. The above postures may be directed towards an individual or towards the group when an animal returns to the group. In the former case they are clearly threat postures and in the latter probably indicate the status of the animal. Rubbing the neck and head on trees may be an indirect means of olfactory communication which indicates the presence of the animal to others. Likewise, defaecation during 'scraping the ground' and 'rubbing the neck' may have the same communicative function.

Sheep

Feral Soay sheep on St. Kilda, outside the breeding season, are to be found divided into male flocks and female flocks, each flock having a home range (Grubb and Jewell, 1966). At the rut the males leave their groups and join the female groups until the rutting is over when the male groups are formed again. In domestic flocks that had little culling, ewe family groups, each with its own home range, have been found. If shepherded out of the home range these sheep were reluctant to feed (Hunter and Milner, 1963). Such organization can cause difficulties in a selection programme for it means that related sheep will all be on one type of habitat and the genotypes will not be randomly distributed with respect to environmental factors.

Within groups, dominance hierarchies may be found when food resources are concentrated, as at a food trough, and rank under these conditions does not appear to be related to body weight (Arnold and Dudzinski, 1978). Different breeds tend to disperse themselves more than others. The mean nearest neighbour distance, for example is 3.4 m for Suffolks and 7.5 m for Blackfaced sheep. Another study revealed that 91% of Merinos were within 8 m of their nearest neighbour, while for lowland and hill breeds the percentages were 84

Fig. 2.20   A Rocky Mountain Bighorn ram. This is a 10-year-old. (From Geist, 1971.)

and 67 respectively. However, within a breed, nearest neighbour distance will depend on the current behaviour. Furthermore the absolute values of nearest neighbour distance during grazing will of course be governed partly by state of pasture and type of vegetation; Dobie (1979) found differences in nearest neighbour distances in Blackfaces grazing in grass or heather in the same flock. Furthermore, interacting with this measure is the tendency of breeds to form groups of difference sizes. Merinos, for example, do not break up into small groups unless there is a shortage of food but will remain in groups of up to several thousand whereas some of our hill breeds form small groups (Arnold and Dudzinski, *loc. cit.*). Finally, the physical features of the environment will also affect the pattern of dispersal. Leadership in sheep, and as we have seen in cattle, does not have the same meaning that is implied in the case of leadership in a wolf pack or baboon troop. Arnold (1977) has stated that it is performed by the less gregarious sheep. On the other hand Syme and Syme (1975) investigated leadership in a small group of sheep subjected to three types of treatments and concluded that leadership in sheep reflects individual differences in reactivity to a number of distinct stimuli, or the same stimuli in contrasting environments. Thus one might expect one sheep to lead in one situation and another in another. The social interactions that govern group movement and coordination are unknown. Due to the absence of an obvious alpha animal which overtly influences movement, such interactions are undoubtedly complex and vary according to context. Little attention has been paid to communication between sheep outside courtship and we know little about the role of communication in the movement and spacing of sheep. Crofton (1958) has claimed that the orientation of sheep relative to one another is such that the nearest neighbour tends to be 110 degrees from the orientation axis of each animal.

It has also been reported that whilst grazing one member of the group will be alert. Furthermore Lawrence and Wood-Gush (1982) found that in a tall brassica crop lambs appeared to be more alert than in a shorter turnip crop where lines of vision were longer. Furthermore Crofton (*loc. cit.*) reported that a leading sheep that cannot see its neighbours raises its head more and grazes less continuously than the others (Fig. 2.21). Likewise in the Bighorn sheep, animals alone or in small groups spend more time scanning the environment than those in larger groups (Berger, 1978). Geist (1971) and Berger (1979) have reported and described 23 displays in the

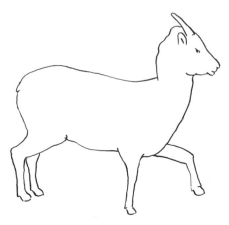

Fig. 2.21 The 'stiff head-erect and ears back' posture of the alarmed Bighorn sheep. (From Geist, 1971.)

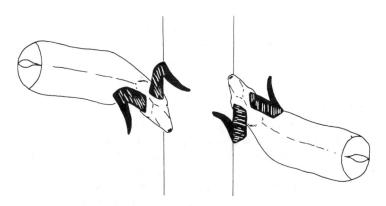

Fig. 2.22 The positioning of the horns and body during a horn display in the Bighorn sheep. (From Geist, 1971.)

Bighorn sheep. Most of these are probably to be found in domestic sheep, for Lawrence and Wood-Gush (1982) in a relatively brief period found 11 of them in Blackface and Suffolk lambs. The ones with communicative value seen by them were described by Geist (*loc. cit.*):

The front kick. Performed with the extended front leg this is done mainly by a dominant ram which hits the other sheep ventrally on the chest, or belly, or haunches, or occasionally on the neck or chin.

The twist (Fig. 2.23). The ram dips his head low, simultaneously rotates it sharply about its long axis, flicks rapidly with the tongue and expels a harsh loud growl. It is sometimes combined with the 'front kick' and this combination has been called the 'nudge' (Banks, 1965). In the Bighorn sheep it appears to be a serious threat and is more common in that species than in the Thinhorn sheep.

Upright attention posture. The sheep suddenly freezes and stares in one direction. The ears are pitched forward and the body may be orientated along the line of sight. It alerts other sheep.

Fig. 2.23   The twist display in the Bighorn sheep. (From Geist, 1971.)

Upright alarm posture. The sheep raises its head and holds it up rigidly while walking with tense steps. This also draws the attention of other sheep.

Kiley (1972) has summarized the little information there is on the vocalizations of sheep and their close relatives. Grunts are given on the greeting of equals and when in close contact; snorts are given when startled and, as is well know, bleats when isolated. Whether there is olfactory communication in sheep is an open question.

The Horse

All equids can interbreed in captivity so that their relationship is very close and, indeed, this is borne out by the fact that a number of behaviour patterns are almost identical. However, there are differences between species in social organization. Studies have been carried out on some zebra species and the wild ass in Africa, by Klingel (1974). In the plains and mountain zebra, the family group – consisting of the stallion, his harem and their young – live in home ranges which are shared sometimes with surplus males living in stallion groups. In these species mating only takes place between a family stallion and his mares and there is no competition amongst stallions for adult mares. Other stallions are sometimes, however, attracted to adolescent mares during oestrus and they try to inveigle the young mare away. Only at about 2 to 2.5 years of age, at her second oestrus, does the mare become a permanent member of a group.

Fig. 2.24 Submissive teeth-clapping by an immature male (left) to a stallion. (From Feist, 1971.)

Klingel (*loc. cit.*) claims that the attraction of the young outside stallions to the young mare is due to the fact that she adopts a conspicuous oestrus stance whereas adult mares are less conspicuous. When a stallion is old his place is taken by a younger male and the group remains otherwise intact. Young stallions may remain with their natal family up to 4 years of age, but may leave earlier. At 5 to 6 years of age they start to compete for young mares.

In the wild ass and Grévys zebra the social organization is very different. They are found alone or in a variety of different associations. In Grévys zebra the male has a very large territory with an average size of 5.75 km² and Klingel's observations suggest that those of the wild ass are even larger. These territorial boundaries are defended only when there is a mare in oestrus, and within the territories the transient males do not interfere with the territorial male. The boundaries are characterized by dung piles and the territory owner also advertises his presence vocally. Klingel suggests that the territorial system aids mating in these species as the presence of an oestrus female outside a territory causes so much fighting that she may not be mated.

In the plains and mountain zebra the stallion is the alpha animal and the mares are ranked in an hierarchy, but in the territorial species Klingel could find no sign of a dominance hierarchy. The group ties in the plains and mountain zebra are very strong; if a member is missing, the group will look for it, except that mares do not look for mares or older colts. When alerted, adults will ensure that any young are

Fig. 2.25 Threat behaviour in the horse. The mare on the left threatens two others which know one another. The central mare which was previously the dominant mare shows some aggression by the position of her tail, her bent neck and carriage but her ears indicate submission. (From Schäfer, 1975.)

awakened, a function which in most ungulates is done by the dam.

In feral populations of the domestic horse which are not culled or cropped the basic unit is a stallion, his harem and their young. Generally these family units occupy a home range but in one feral population on an island off the coast of the USA the males are reported to be territorial. In the New Forest ponies (Tyler, 1972) the same basic structure is found although the stallions are generally removed by man. Domestic mares, like the New Forest ponies, form a hierarchy (Houpt *et al.*, 1978; Francis-Smith, 1979). In the group studied by Francis-Smith the distance between mares was generally related to their proximity in the social hierarchy. These horses also showed a high degree of synchrony in their behaviour. Agonistic behaviour was restricted to threats and submission, and the threats were mostly threats to bite, rather than threats to kick and consisted of a slight inclination of the head with the ears back. Submission consisted of a horse taking a few steps back. Mutual grooming was seen in this group and was always started by the more dominant animal of the pair.

CONCLUSIONS

From the work of Wilson and others, it is apparent that to understand the social behaviour of wild species we must at least know something about the demography, spacing behaviour and intra-group behaviour of the population under scrutiny. In farm animals, mostly kept in monocaste groups of a single sex, the use of these parameters is not so obvious. The main research effort dealing with the social behaviour of these species has concentrated on social hierarchies to the exclusion of other factors. Furthermore, in these studies the emphasis to date has been on dominance, while submission has been neglected, although in many cases it may be more important in maintaining relative rankings than dominance. It would certainly be advantageous if attention were paid to the natural grouping of animals and to their communication. Although our farm animals are the descendants of social species we tend to keep them in large groups for which their evolutionary history may not have adapted them. We therefore need to have studies in which they can have groups and social spacing of their own making, so that we can provide better housing.

POINTS FOR DISCUSSION

1. Describe how the early social experience of a farm animal can affect its economic performance.
2. Discuss how her social behaviour could affect the performance of (a) a dairy cow and (b) a hill ewe.
3. List the forms of communication used by one species of farm livestock and describe the contexts in which they might be used.
4. Watch the behavioural responses of a farm animal to a stockman and compare them with those used to conspecifics.

FURTHER READING

Arnold, G.W., Wallace, S.R. and Rea, W.A. (1981), Associations between individuals and home-range behaviour in natural flocks of three breeds of domestic sheep, *Appl. Anim. Ethol.*, 7, 238–258.

Barrett, P. and Bateson, P. (1978), The development of play in cats, *Behaviour*, **66**, 106 120.

Beilharz, R.G. and Mylrea, P.J. (1963), Social position and behaviour of dairy heifers in yards, *Anim. Behav.*, 11, 522–533.

Bosc, M.J., Bouissou, M.F. and Signoret, J.P. (1968), Consequences de la hierarchie sociale sur le comportement alimentaire des bovins domestiques, 93me Congrés Nationale des Societes Savantes Tours, *Sciences*, 2, 511–512.

Donaldson, S.L., Albright, J.L. and Black, W.C. (1972), Primary social relationships and cattle behaviour, *Proc. Indiana Acad. Sci.*, 81, 345–351.

Ewbank, R. (1976), Social hierarchy in suckling and fattening pigs: A review. *Livestock Prod. Sci.*, 3, 363 372.

Hughes, B.O. and Wood-Gush, D.G.M. (1977), Agonistic behaviour in domestic hens: The influence of housing method and group size, *Anim. Behav.*, 25, 1056–1062.

Hunter, R.F. (1964), Home range behaviour in hill sheep, in *Grazing in terrestrial and marine environments* (Crisp, D.J., ed.), Blackwells Scientific Publications, Oxford, pp. 155–172.

Komai, T., Craig, J.V. and Wearden, S. (1959), Heritability and repeatability and social aggressiveness in the domestic chicken, *Poult. Sci.*, **38**, 356–359.

Leaver, J.D. and Yarrow, N.H. (1980), A note on the effect of social rank on the feeding behaviour of young cattle on self-feed maize silage, *Anim. Prod.*, 30, 303–306.

McBride, G. (1958), Relationship between agressiveness and egg production in the domestic hen, *Nature*, 181, 858.

Syme, G.J. and Syme, L.A. (1978), *Social Structure in Farm Animals*. Elsevier Scientific Publishing Co., Oxford.

3
FEEDING BEHAVIOUR

INTRODUCTION

When analysing behaviour such as feeding behaviour we wish to know the physiological factors controlling it as well as the external factors involved. Once the internal factors become operative the animal is likely to show appetitive behaviour in which it will be searching for certain objects. When these have been found the appetitive behaviour will give way to consummatory behaviour. Diagrammatically the system may be shown as follows:

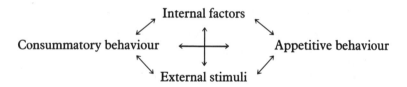

The two-way arrows indicate that interactions occur both ways; once appetitive behaviour has begun it will affect the internal state of the animal and, as we shall see later, the internal state of the animal affects its perception of external factors and so on.

INTERNAL MECHANISMS CONTROLLING FEEDING BEHAVIOUR

Hunger is a commonly used term but its complexity is soon appreciated when we try to measure it. Commonsense would say that an animal that is eating a great deal is very hungry. However, if we deprive rats or fowls of food for various lengths of time, the relationship between the length of the starvation period and the amount of food eaten subsequently is seen to be very unreliable. In chickens the amount eaten after 48 hours of food deprivation is no greater than after 24 hours of food deprivation. One might then say: what about the rate of feeding? A hungry animal will eat more quickly than a non-hungry one, but in fowls the rate of feeding is also not very reliable, for again there is no difference between a fowl's pecking rate after 24 and 48 hours food deprivation.

Miller and his coworkers (Miller, 1957) compared the effects of several hours of food deprivation on four different measures of hunger in the rat:

1. Volume of enriched milk required to satiate rats.
2. The amount of quinine adulteration needed to stop the rats from feeding.
3. The amount of bar pressing to get food on a variable reward schedule.
4. The rate of stomach contractions after various hours of deprivation.

As shown in Fig. 3.1, the different measures respond differently to increasing food deprivation. There is very little correlation between

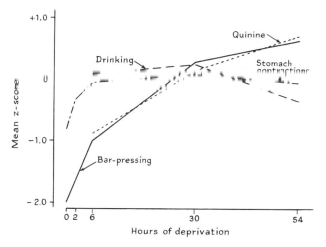

Fig. 3.1 Comparisons of 4 measures of hunger: (a) drinking the volume of enriched milk required to satiate the rats: (b) the amount of adulteration with quinine required to prevent eating: (c) the rate of bar pressing and (d) the sum of the excursions of the record of stomach contractions measured from an implanted balloon. (From Miller, 1957.)

them, and these findings emphasize the need for more than one measure when trying to gauge the strength of the internal state of a behavioural tendency.

The central nervous system

In 1904 it was reported that the hypothalamus was involved in certain cases of obesity, but it was not until 1940 that Hetherington and Ranson claimed an exact site in the hypothalamus, the ventro-medial nucleus. Rats with bilateral lesions of this nucleus showed hyperphagia (over-eating) and became obese. It was also found that lesions in the lateral hypothalamus led to the converse; the rats ceased to eat (became aphagic) and could only be cajoled back to feeding by being

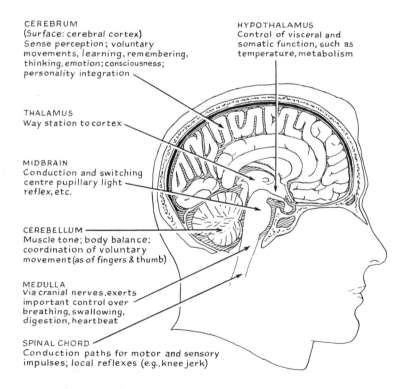

CEREBRUM
(Surface: cerebral cortex)
Sense perception; voluntary
movements, learning, remembering,
thinking, emotion; consciousness;
personality integration

HYPOTHALAMUS
Control of visceral and
somatic function, such as
temperature, metabolism

THALAMUS
Way station to cortex

MIDBRAIN
Conduction and switching
centre pupillary light
reflex, etc.

CEREBELLUM
Muscle tone; body balance;
coordination of voluntary
movement (as of fingers & thumb)

MEDULLA
Via cranial nerves, exerts
important control over
breathing, swallowing,
digestion, heartbeat

SPINAL CHORD
Conduction paths for motor and sensory
impulses; local reflexes (e.g., knee jerk)

Fig. 3.2 The main sub-divisions of the central nervous system and their functions. (From Hilgard, 1957; after Best and Taylor, 1955.)

given liquid diets which were gradually thickened over a span of time. Conversely, electrical stimulation of the ventro-medial nucleus caused hungry rats to stop feeding while stimulation of the lateral hypothalamus caused satiated rats to feed.

These findings resulted in the postulation (e.g. Stellar, 1954) that there was a hunger centre in the lateral hypothalamus and a satiety centre in the ventro-medial hypothalamus. It was further postulated (e.g. Brobeck, 1957) that these centres were activated by changes in the blood. Three possible means of activation, not necessarily mutually exclusive, were suggested:

1. A downward change in the level of blood glucose would activate the 'hunger' centre and an upward rise, the 'satiety' centre.
2. Similar changes in the levels of circulating lipids would have similar effects.
3. Changes in the temperature of fluids reaching the hypothalamus would have these effects (Bray, 1976).

This concept is now seen to be an over simplification of the control of feeding behaviour (e.g. Grossman, 1975; Rolls, 1976). Many other parts of the brain have been shown to be involved including the cortex, the basal ganglia, the limbic system (Fig. 3.3) and structures in the medulla oblongata (Bell, 1971). Figure 3.4 gives some indication of the factors involved in the control of feeding.

As we all know from experience the smell of favourite food will motivate us to feel hungry even if up to that point we had not consciously felt hungry. By the use of recording from single cells it has been found, in two species of monkey, that there are neurones in the lateral hypothalamus which change their firing rate when the hungry monkey sees food. Such cells do not change their firing rate when the monkey sees non-food items nor when the monkey feeds in the dark; they are specific to visual cues emanating from food or food sources. Others have been found which change their firing rate when the food is tasted, while yet another category responds to both the sight and taste of food. All these however only respond when the monkey is hungry. Learning can affect the responses of these neurones in the monkey: if they have been given glucose from a black syringe, the sight of the syringe will evoke the change in the cells while the sight of a white syringe from which they have received physiological saline will not. It seems that the role of these cells in feeding behaviour is to guide and maintain the behaviour and while hunger must be present for

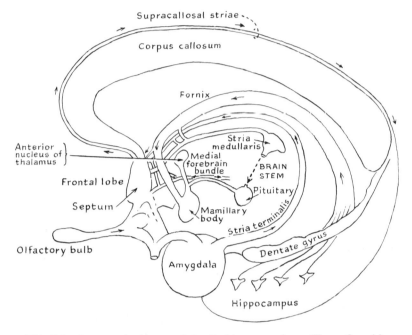

Fig. 3.3 A general picture of the limbic connections. (From Smythies, 1970.)

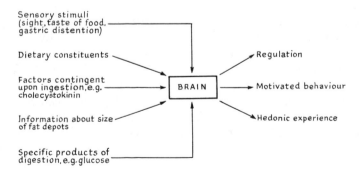

Fig. 3.4 Factors involved in the control of feeding. (From Blundell, 1976.)

these cells to function, it acts as a gating mechanism, and there is no evidence that these mechanisms start or stop feeding (Rolls, 1976).

It has been sugested that the ventro-medial nucleus may normally serve to control a number of autonomic and endocrine reflexes that are concerned with the metabolism of food and which are triggered off by sensory contact with foodstuffs. Some of these reflexes function to prepare the gastro-intestinal tract to move, digest and prepare the viscera to metabolize and store nutrients. They also allow the animal to monitor its intake, while some will have inhibitory effects as well; lesions in this nucleus are suggested to cause hyperphagia and obesity by disrupting this control (Powley, 1977). This is an attractive hypothesis in that it helps to explain the behavioural differences between rats with these lesions and normal hungry rats, but as yet is mainly speculative.

Feed-back mechanisms

Once food has been ingested, signals from various sorts of receptors will be relayed to the brain. The action of these mechanisms, known as feed-back mechanisms, will affect the state of hunger as revealed by the feeding behaviour of the animal, or in ourselves affect our subjective feelings of hunger or satiation.

Sensory stimuli derived from the mouth and intestinal tract (pre-gastric and gastric factors) are important in controlling the amount of feeding behaviour shown. Experiments were carried out on single rats in a Skinner box as described in Chapter 1. Using a fixed schedule of reinforcement, the rat received a set amount of enriched food for every response. It was found that rats that had been allowed to drink 14 cm^3 of enriched milk just before testing, pressed the lever less than rats that had had 14 cm^3 of enriched milk introduced into their stomachs by a tube, and that this second group of rats showed less lever pressing than rats that had had 14 cm^3 of physiological saline introduced directly into the stomach by a tube. Exactly the same results were obtained in another experiment in which the behavioural measure was the amount of enriched milk drunk after the treatments instead of the number of times the lever was pressed. The effects were immediate so that digestion of the milk played no role (Miller, 1957). Stimuli from the mouth and upper part of the intestinal tract therefore appear to be important and it has also been found in mice that the taste of the food will affect the length of the period of feeding, bitter food decreasing it

(Wiepkema, 1971). Pre-gastric factors have also been found to be important in the dog in controlling the precise regulation of food intake, for dogs given their normal daily intake of food intra-gastrically will eat extra food orally if it is offered and will only reject extra food after having 50% more than their normal intake intra-gastrically (Bray, 1976).

The importance of stomach contractions were tested in a very old experiment carried out in 1912 by Cannon and Washburn. They demonstrated that, in man, stomach contractions occurred concurrently with subjective feelings of hunger and suggested that the state of distension of the stomach might be the signal for the subject to start or stop feeding. However it was found that this was too simplistic; for example a person without a stomach can feel hungry, and today gastric distension is considered to be one of the several feedback mechanisms that influences feeding behaviour. In pigs, for example, if 3 to 12 day old piglets deprived of milk are given a gastric loading of saline, a non-nutritive substance, they will ingest significantly less milk from an artificial teat than piglets that have been equally deprived of milk, although it is not as effective in suppressing suckling as milk or glucose (Stephens, 1975). Gut distension not only increases the firing rate of neurones in the ventro-medial nucleus but also increases the firing rate of the vagus nerve. If this nerve is sectioned in the rat, the animal reduces the size of individual meals but

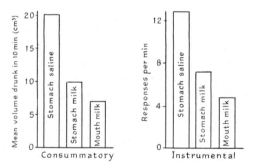

Fig. 3.5   An injection of 14 cm³ of enriched milk directly into the stomach reduces hunger more than does the same volume of isotonic saline but milk drunk by the mouth prduces still greater reduction. The same pattern of results is shown by the volume of milk drunk or by the amount of bar pressing. (From Miller, 1957.)

increases the frequency of feeding and it is possible that the vagus plays a role in relaying the state of distension of the gut to the brain (Rolls, 1976).

Evidence from parabiotic rats, which have been united surgically to share a common blood system, suggests that gastro-intestinal hormones may be effective in controlling feeding behaviour. If one of these animals is allowed to feed to satiation and one hour later the second animal is given access to food, it is found to eat less than expected, suggesting that some circulating factor is transferred from one rat to the other (Bray, 1976). One such factor is the hormone cholecystokinin (CCK) which is produced by the intestine. Gibbs *et al.* (1973) studied the feeding behaviour of rats after they had received intraperitoneal injections of either CCK or saline. CCK led to a significant decrease in food intake which could not be attributed to the animals feeling sick. Furthermore the substance had no effect on water intake.

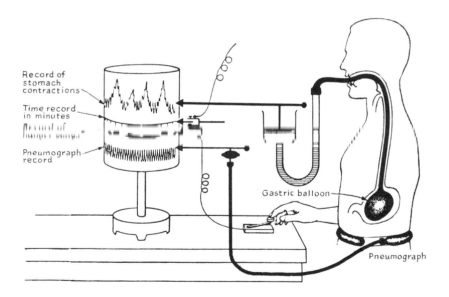

Fig. 3.6 An arrangement for correlating hunger pangs with stomach contractions. Note that the reported hunger pangs correspond closely to the periods when stomach contractions are at their maximum. (From Hilgard, 1957; after Cannon, 1934.)

Plasma glucose levels are still considered to be important controlling factors and, in addition to gluco-receptors in the ventro-medial and lateral hypothalamus (Blundell, 1976), hepatic gluco-receptors and possibly some in the duodenum may be involved (Bray, *loc. cit.*). However, some workers dispute the importance of gluco-receptors in normal feeding, suggesting that they are only called into play when the glucose levels are very low (Strubbe *et al.*, 1977). Insulin, which plays a role in glucose metabolism, may also be involved. Injections of insulin lower blood sugar, and the concentration of insulin in the plasma drops to a low level just before a meal. Its function in feeding behaviour is unclear (Strubbe and Mein, 1977).

With regard to the lipostatic hypothesis, glycerol has been suggested as a likely metabolite from fat which could control feeding behaviour. It is released during the hydrolysis of triglycerides in adipose tissue and cannot be re-utilized at the same site because a key enzyme, glycerol kinase, is either absent or in very low concentration in adipose tissue. Thus glycerol is released into the circulation in proportion to the hydrolysis of triglycerides. It is transported to the liver where it can readily be converted into glucose (Bray, *loc. cit.*). Thermostatic control is suggested largely because environmental temperature is an important modulator of feeding behaviour, but how it operates in detail is not known. Amino acids as controlling mechanisms in specific cases will be discussed in detail when specific hungers are considered, but their role in regulating feeding in other circumstances is still vague.

An energostatic control mechanism has been proposed (Booth, 1972). Rats given stomach loads of metabolizable substances and then allowed free access to food for one to two hours, later show deficits in food intake relative to rats given stomach loads of saline or urea. The observed decrease in food intake is correlated with the expected energy yield of the load showing that the inhibiting effects are energostatic rather than glucose specific.

Certain monoamines have also been suggested as factors controlling feeding behaviour. These include noradrenalin, dopamine and serotonin, and are found in certain nerve cells, in the cell bodies, axons and terminals. Noradrenalin has been found in cells in the medulla oblongata, midbrain, hypothalamus including the VMH (ventro-medial hypothalamus) and LH (lateral hypothalamus), thalamus, hippocampus, cortex and cerebellum. The pathways connecting some of these areas have been worked out. Seven distinct pathways for dopamine have been found, one of which passes through the lateral hypothalamus. Using a certain neurotoxin it is possible to cause degeneration of neurones that contain either noradrenalin or dopamine by injecting it into a particular area. Using this technique involving the LH nucleus it is possible to produce rats that resemble rats with bilateral LH lesions. Such experiments have shown that the aphagia and adipsia are seen only after extensive destruction of the dopamine neurones, suggesting that these impairments are due to more general damage rather than to the destruction of a feeding centre.

Mono-amino neurones may be involved in the hyperphagia seen after VMH damage. However the behavioural changes seen after the destruction of the cells containing noradrenalin do not precisely parallel those seen after VMH lesions, for the hyperphagia and weight gain are less pronounced. From various sources of evidence it seems that dopamine may be involved in hunger and noradrenalin in satiety, but the involvement of these mono-amines may be indirect in that they involve more general functions (Marshall, 1976; Blundell, 1976). Serotonin levels in the brain are directly affected by food intake, for following a meal containing carbohydrate, but lacking protein, serotonin synthesis in the brain is accelerated and its levels raised. It has been suggested therefore that serotonergic neurones might participate in the control of calorie or protein consumption over the short or long term (Blundell, *loc. cit.*).

In adult ruminants the gastric factors that control short-term feeding behaviour are different from those in monogastric animals. The main products of fermentation in ruminants are acetates, propionates, butyrates and lactates and these alone or in some combination may act upon the rumen to affect feeding behaviour. Other factors may be rumen distention, the pH of the rumen fluid and its osmolarity (Baille and Forbes, 1974).

One method of trying to assess how short term control of feeding is affected has been to study the pattern of meals and the intervals between meals. In most animals studied (rat, fowl, Japanese quail and cow) there is a significant correlation between meal size and the length of the *succeeding* interval to the next meal, i.e. a small meal is followed by a short interval and a large meal by a long interval. On the other hand there are not such strong correlations between meal size and the *preceding* interval. A long interval may be followed by a meal of the same size as that taken after a short interval.

The correlation between meal size and succeeding (post-prandial) interval suggests that the trigger mechanism involves a monitoring of a critical level *below* which ingested food should not fall, i.e. after a small meal this level is reached more quickly than after a large meal. This level may involve a physical parameter or a biochemical one, or both, and fit in with the theories which have been outlined.

APPETITIVE BEHAVIOUR

Hungry laboratory rats show increased locomotor activity and are more reactive to changes in the environment than non-hungry ones. They become much more active when lights are switched on than do non-hungry rats. This increased activity would correspond to an animal in the wild searching for food in its territory. Generally appetitive behaviour is considered to be flexible and susceptible to change through learning. In the case of animals kept intensively, or semi-intensively, elements of learning will be found readily in their appetitive behaviour; e.g. they will tend to gather around the usual point of feeding at times when food is to be given.

EXTERNAL FACTORS CONTROLLING FEEDING BEHAVIOUR

One might imagine that the searching animal carries a picture of its goal and recognizes it by its entirety. However, in cases in which detailed analysis has been conducted, we find that this is not so; the hungry animal responds only to certain stimuli emanating from the goal. The hungry herring gull chick provides us with one of the best examples. The young chicks solicit food by pecking at the bill of the parent, which in this species is yellow with a red patch on the lower mandible. Tests were made (Fig. 3.7 and Tinbergen and Perdeck, 1950) with a number of cardboard models in which the experimenters varied the head colour, head shape, bill colour, colour of the patch on the lower mandible and the degree of contrast between the patch and the bill. Chicks that had never seen a model or a real gull's head were used and all variables were systematically controlled. It was found that the mandible patch was the most important feature of the head in eliciting the pecking response: not only was its red colour important but also the degree of contrast with the background colour of the bill. The colour or shape of the head beyond the bill were unimportant.

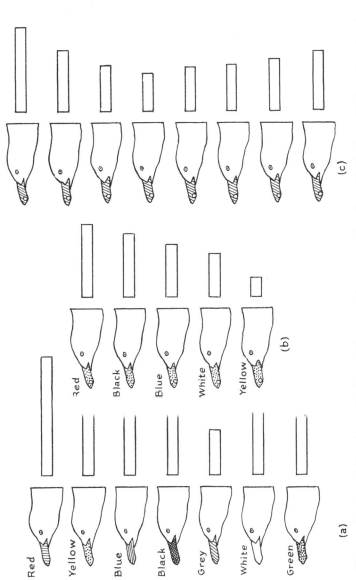

Fig. 3.7 Three series of model heads used in the pecking tests (a) measures the effect of bill colour; (b) that of patch colour (all bills were yellow) and (c) the effect of varying the contrast between patch and bill colour. The length of the bar next to each model is proportional to the number of pecks given to it. (From Tinbergen and Perdeck, 1950.)

The red spot is spoken of as a releaser for the begging or gaping response in the herring gull chick, and it is postulated that it interacts with an innate releasing mechanism which then activates the relevant behaviour which may, in some cases, be relatively simple. In other cases, in which the stimulus has not evolved primarily for its function as an eliciting stimulus, it is known as a sign stimulus. For example, the speckling on the herring gull's egg stimulates the bird to sit on it, but it is likely that speckling evolved primarily as a form of camouflage.

Sometimes more than one sign stimulus is involved, and although one might suffice to elicit the response, the others are additive in their effect. The term heterogeneous summation is used to describe these cases.

The sign stimuli or releaser for feeding behaviour have not been investigated much in mammals. In new born calves, one of the sign stimuli in teat seeking is apparently a recess. In beef cows the teats are more easily found by newly-born calves than in dairy cows, because the conformation of the beef cow is such that the udder is at the part of the under-belly with a clear-cut recess between the abdomen and the udder, whereas in the multiparous dairy cow the udder generally slopes down from the abdomen (Fig. 3.8). In the latter case the calves often end up by searching round the forelegs (Selman *et al.*, 1970). In the case of the foal one would expect that some stimulus from grass would elicit grazing, but although they have a propensity to chew it is necessary for them to be with other grazing equids, otherwise they direct their chewing to extraneous objects rather than to the grass (Glendenning, 1977). In young domestic chicks certain colours release pecking more than others and certain shapes are preferred to others (Wood-Gush, 1971).

Feeding behaviour can also be released by social factors. Allelomimetic behaviour has been defined as two or more animals doing the same thing at the same time with some degree of mutual stimulation. Social facilitation is defined as any increment in the behaviour resulting from the presence of the other animal or animals, e.g. Tolman and Wilson (1965) have investigated the effect of companions on the feeding behaviour of young chicks. The tests were started when the chicks were 5 days old. When not tested, the chicks were kept in large groups. They were tested for the amount eaten (a) in isolation, (b) in pairs, (c) in pairs separated by a transparent division and (d) in groups of 4 and 16. The presence of one companion increased feeding, but increasing the number of companions beyond one did not have any further effect. However the presence of the transparent division decreased feeding, apparently by making the chicks more frightened of the situation when they found that they were unable to join the other chick. Arnold (1970) reports that social facilitation may increase the intake of the grazing ruminant by 20%.

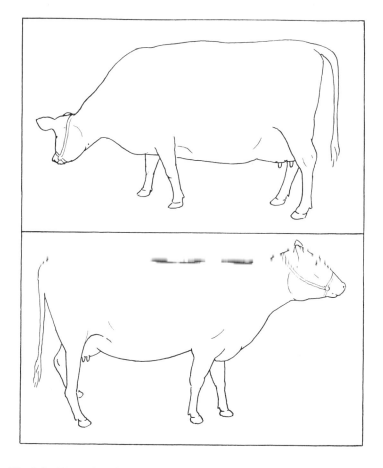

Fig. 3.8   Examples of (top) a cow with a poorly shaped udder for the calf and (bottom) a cow with a good shape for the calf. (From Selman *et al.*, 1970.)

CONSUMMATORY BEHAVIOUR

This part of the behaviour pattern was originally thought by ethologists as being fixed and innate, but it has been found to be much more flexible than previously thought. A good experiment by Cruze in the 1930s, with chicks, is an example. By keeping young birds in the dark for varying times from hatching until testing, and by giving some of them controlled amounts of practice he was able to show that, with training, there is considerable improvement in the accuracy of pecking. However it is also apparent that maturation also plays a role in the absence of practice (see Table 3.1).

Some striking adaptations of feeding behaviour have taken place, some involving modification of the chain of behaviour near the point of ingestion, while others concern the appetitive behaviour. One interesting case of the former type arose in a troop of Macaque monkeys in Japan in which one female 'invented' the washing of sweet potatoes before eating them and this behaviour became established in the troop (Kawamara, 1962). The use of tools for gathering items of food or for facilitating its availability occurs in several species, e.g. chimpanzees use stalks of grass to collect termites (Goodall, 1964) and several species of birds utilize extraneous objects to assist them in getting food. Thrushes use stones as 'anvils' to break open snail shells (Sheppard, 1951), the Egyptian vulture throws stones from a standing position at ostrich eggs to break their tough shells, and in the Galapagos islands the finch, *Cactospiza pallida*, uses cactus spikes on twigs to prise insects out of crevices (Fig. 3.9). Detailed observations

Table 3.1   The pecking accuracy of chicks at different ages before and after 12 hours of practice. Each figure represents an average from 25 chicks. (From Manning, 1979, after Cruze, 1935.)

Age(h)	Practice (h)	Average misses (25 pecks)
24	0	6.04
48	12	1.96
48	0	4.32
72	12	1.76
72	0	3.00
96	12	0.76
96	0	1.88
120	12	0.16
120	0	1.00

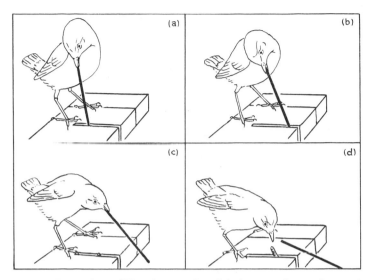

Fig. 3.9 Cactospiza pallida using a tool to dislodge a mealworm from a slot. Note that the finch moves the tool in the direction of the open end of the slot. (Millikan and Bowman, 1973.)

on juvenile *C. pallida* have been carried out by Millikan and Bowman (1973). Under the experimental conditions the birds would only use a twig to find when the food (a meal worm) could not be picked out with the bill, and then the birds tended to break off a twig branch rather than pick one off the ground. Sometimes a bird picked a twig when no food was in sight and at times carried them to cracks some distance away. The actual probing movements were adjusted to the type of tool being used. Long flexible probes such as long pine needles, for example, were sometimes held in the middle rather than at one end (the most common method). Long probes were sometimes broken into two and on one occasion a probe that was too long for the bird to use standing was utilized successfully by the bird on wing. However the study was unable to show how the behaviour develops and we have little insight into how tool usage arose in this species.

SPECIFIC HUNGERS

A specific hunger is demonstrated when an animal is deficient in one particular nutrient, and apparently recognizes that fact by its ability to choose the missing nutrient when given a choice between it and other nutrients.

Much of the published evidence on specific hungers is equivocal. Man is apparently able to form a specific hunger for salt (cited by Denton, 1967) but man is notoriously bad at balancing his diet, for in our history malnutrition has been rife when the proper food was plentiful (Drummond and Wilbraham, 1957). There are many papers published, mainly in the 1930s, on the ability of poultry to balance their rations when given a free choice. However the foodstuffs offered were natural ones containing a large number of nutrients, so that a random selection would probably have led to the consumption of an adequate diet. Likewise a very widely quoted experiment claimed to show that human babies are able to select a good diet, but a random choice of the foods offered would have given a baby an adequate diet and, of course, no detrimental items were offered (Davies, 1928). The evidence from sheep is not too convincing either. Weir and Torrel (1959) reported from studies on sheep with a ruminal fistula that the sheep were consistently selecting forage higher in protein and lower in crude fibre than necessary. Tribe (1950) reported that sheep fed on a self choice system which included linseed cake meal, selected a diet with approximately twice as much protein as necessary. However when the linseed was replaced by fish meal, less protein than required was eaten. Clearly the perceptual qualities of the protein source are more important to the sheep than its physiological needs.

In another experiment Gordon and Tribe (1951) allowed individually housed ewes a free choice of protein and carbohydrate sources, roughages, minerals, and salt from conception until lambing. Their protein intake was too low and they over-ate during early pregnancy and did not eat enough in the later stages of pregnancy. Arnold (1970) has reported that sheep also graze according to convenience under certain conditions. In Australia, sheep on dry herbage may not select the legume component at a time when it would be nutritionally advantageous because it is in the least accessible position in the plant canopy. It appears that rats can generally balance their ration in rigorous selection experiments, but there is individual variation, and proper protein selection depends a great deal on the perceptual properties of the protein.

Animals suffering from a particular deficiency will show exaggerated appetitive behaviour, often sampling bizarre objects as food. Denton (1967) cites some cases and Wood-Gush and Kare (1966) showed that chickens deprived of calcium showed more exploration of a strange arena, and pecked at novel objects in it, significantly more than control birds. Some specific nutritional deficiencies have been correlated with outbreaks of feather pecking and cannibalism in fowls (see Wood-Gush, 1971 for reviews). Such exaggerated appetitive behaviour should serve to bring the animal into contact with the needed substance but how does it recognize that substance and what is the underlying control?

Sodium appetite

It has been shown by many workers that if the adrenal glands are removed from rats, these animals will show an immediate appetite for salt (Rozin, 1976). In a situation in which they are given a choice between water and NaCl solution they will choose the NaCl even when the concentration of salt is so low that the controls are showing no choice between the NaCl solution and pure water. Also this preference is very specific to sodium salts and not to other minerals, although the rat may generalize to lithium chloride or potassium chloride. If deoxycorticosterone is implanted into these rats their NaCl intake falls back to normal (Denton, 1967).

Selection mechanisms

How does the rat recognize the nutrient? At one time it was thought that this preference for NaCl was accompanied by a heightened sensitivity of the taste buds for NaCl. However Pfaffman and Bare (1950) exposed the chorda tympani, which is a branch of the facial nerve, and made direct recordings from the nerve fibres while stimulating the tip of the tongue with NaCl solutions of various strengths (Fig. 3.10). They used both adrenalectomized rats and control rats and found no differences in the minimum concentration of the solutions to which the nerve fibres of the two classes of rats would respond. Both responded to solutions of about 0.008% NaCl, thus demonstrating that normal rats were registering the presence of salt but not acting on the information, whereas in the adrenalectomized rats the preference and sensory thresholds more or less coincided. An experiment showing that normal rats can act upon this sensory information, was one in which the rats were given a choice

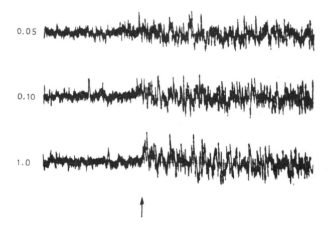

Fig. 3.10   Three oscillograph records of the responses in the chorda tympani nerve produced by sodium chloride solution on the tongue. The stimulus intensity in percentage concentrations is shown at the left and the beginning of the nerve response is indicated by the arrow. The strip of record shown covers a 1/10 second interval. (From Pfaffman and Bare, 1950.)

between water and a 0.009% NaCl solution and given an electric shock if they chose water. In this experiment normal rats very soon showed that they could distinguish between the two solutions.

Sheep which have a parotid fistula also become very depleted of sodium and will select a sodium solution in a choice situation. However their method of selection is very different from rats. While rats whose olfactory lobes have been removed can easily detect the NaCl solutions, sheep whose taste buds have been denervated are able to select sodium bicarbonate (Denton, 1967).

Rats and goats have been taught to press panels or levers to obtain sodium salts, thus suggesting that the animal recognizes the benefit of the salt in some way, for under such conditions learning, as we saw earlier, will occur if the response is rewarded. Thus it is suggested that the mechanism responsible for a specific hunger is a need-reduction mechanism, i.e. the characteristics of the required mineral become associated with a feeling of well-being. Rozin (1967), however, suggested that the animal shows an aversion to the deficient diet, rather than a specific preference for the adequate diet.

Calcium appetite

A specific hunger for calcium has been shown in the rat and fowl (which doesn't show a specific hunger for sodium). Rats which have had their parathyroid glands removed form a specific hunger for calcium and will choose it when given a choice between two solutions (Richter and Eckert, 1937). They will generalize this preference to other calcium salts, e.g. calcium lactate, calcium acetate, calcium gluconate and calcium nitrate, but not calcium phosphate. Also some of these rats show a preference for strontium salts. If the parathyroid gland from another animal is transplanted into the eyeball (thus avoiding immunological rejection), the calcium appetite returns to normal. Rats deficient in calcium show a strong preference for novel diets even if these are calcium deficient.

Fowls deprived of calcium choose a calcium enriched diet in a choice situation even if that diet is fairly unpalatable. However if the calcium salt is too concentrated they will ignore it entirely. The calcium appetite is not immediate as is the sodium appetite of the adrenalectomized rats; it is apparently learnt rather slowly over a period of days (Hughes and Wood-Gush, 1971).

The question of how the calcium is recognized has only been investigated in the fowl and it appears that the fowl will use either taste or visual cues (Hughes and Wood-Gush *loc. cit.*). The rat probably relies on taste cues.

Other specific hungers

Rats deficient in thiamine (B1) form a specific hunger relying on taste cues to do so, although visual and olfactory cues may be used. Likewise for riboflavin (B2), but in the case of pantothenic acid (B6) the results are less clear-cut (Scott and Quint, 1946). However rats will avoid thiamine in solution even if it will benefit them (Rozin, 1976). On the other hand, thiamine deficient rats show an interest in novel diets which is unusual in rats. It is largely this work on thiamine deficiency in rats that has developed the hypothesis suggesting that the underlying motivation for a specific hunger is the avoidance of the deficient diet (Rozin, 1967).

The experiments on the protein selection by sheep as we discussed earlier showed that sheep are not too good at balancing their diets for protein. Likewise rats are not very good at this task. Generally there is a fall in food intake if the protein content is too high or too low. Moreover, rats avoid a diet lacking in one essential amino acid, but if the missing amino acid is added they begin to feed, indicating that the diet was otherwise acceptable. In one such choice experiment, one diet was protein-free and the other lacked tryptophan or threonine. The rats chose the protein-free diet and some even died. However, when threonine was added to the other diet the remaining rats changed over to it.

In one experiment (Leung *et al.*, 1968) rats were fed a diet lacking in either threonine or isoleucine. Their food intake fell but if given minute amounts of the missing amino acid into the carotid artery their appetite recovered. The same amounts injected into the jugular vein (so that the amino acid was diffused through the body) had no effect on their depressed appetites. Therefore it is suggested that within the brain is an area sensitive to the concentrations of essential amino acids in the blood. The results of this experiment pose some very interesting questions about the ending of a meal, if its findings are generally applicable. For example, a farm animal may have a diet with an imbalance of essential amino acids which would cause it to stop feeding before it had eaten enough. A superfluity of these acids might also cause it to stop before it had enough total protein. Furthermore it is possible that total protein requirements might be met before that for some individual amino acids (Harper, 1976).

CONCLUSIONS IN RELATION TO FARM LIVESTOCK

In this chapter we have considered how physiological, social and physical factors control food intake. Here we will consider the relative importance of these with regard to animals kept under modern intensive and extensive conditions. Under intensive conditions with *ad lib.* food in front of them, as say a hen in a battery cage, it is unlikely that their feeding behaviour will be under the control of the physiological mechanisms to any large extent. It is more likely that social factors will predominate and that, in the dull environments of many animals, feeding or toying with the food will have an additional recreational function. Likewise cattle on artificial pastures will have adequate food permanently at their feet and physiological mechanisms probably play a minor role in initiating feeding compared with social and climatic factors. The monotony of diets may also be a factor for consideration in cases in which rapid growth is required. In their evolutionary history many ungulates will have had varied diets which changed with the seasons and with any migratory patterns of the species. All these factors are in need of consideration and investigation if we are to understand the factors controlling food intake in our farm animals.

POINTS FOR DISCUSSION

Grazing behaviour

1. What are the internal factors that initiate grazing?
2. What are the external factors that may initiate grazing?
3. What factors stop grazing?
4. What are the main social factors affecting grazing?
5. Can you think of any evidence of specific hungers in farm livestock kept extensively?

Feeding behaviour under intensive conditions

1. Discuss the importance of social factors in feeding behaviour under these conditions.
2. What is meant by palatability?
3. Turkey poults often do not feed in modern intensive production units. What might cause this aphagia?

FURTHER READING

Grazing behaviour

Arnold, G.W. and Dudzinski, M.L. (1978), *Ethology of free-ranging domestic animals*. Elsevier, Oxford, pp. 97–123.

Broom, D.M. (1981), *Biology of Behaviour. Mechanisms, functions and applications*. Cambridge University Press, London, pp. 150–156.

Duncan, P. (1980), Time budgets of Carmague horses. 11: Time budgets of adult horses and weaned sub-adults, *Behaviour*, 72, 28–49.

Hunter, R.F. and Milner, C. (1963), The Behaviour of Individual Related and Groups of South Country Cheviot Hill Sheep, *Anim. Behav.*, 11, 507–513.

Key, C. and McIver, R.M. (1980), The effects of maternal influence on sheep: breed differences in grazing, resting and courtship behaviour. *Appl. Anim. Ethol.*, 6, 35–48.

Kiley-Worthington, M. (1977), A Review of Grazing Behaviour, in *Behavioural Problems of Farm Animals*. Oriel Press, London.

Odberg, F. and Francis-Smith, K. (1976), A Study of Eliminative and Grazing Behaviour. The use of the field by captive horses. *Equine Vet. J.*, **8**, 147–149.

The relevant sections in the chapters on cattle, horses, pigs and sheep in *The Behaviour of Domestic Animals*, 3rd edn (Hafez, E.S.E., ed.). Balliere, Tindall and Cassell, London.

Feeding behaviour under intensive conditions and its control

Baile, C.A. and Della-Fera, M.A. (1981), Nature of hunger and satiety control systems in ruminants, *J. Dairy Sci.*, **64**, 1140–1152.

Baile, C.A. and Forbes, J.M. (1974), Control of feed-intake and regulation of energy balance in ruminants, *Physiol. Rev.*, **54**, 161–214.

Duncan, I.J.H., Horne, A.R., Hughes, B.O. and Wood-Gush, D.G.M. (1970), The pattern of food intake on Brown Leghorn hens as recorded in a Skinner box, *Anim. Behav.*, **18**, 245–255.

Grovum, W.L. (1979), Factors affecting the voluntary intake of food by sheep. 2: The role of distention and tactile input from compartments of the stomach, *Br. J. Nutr.*, **42**, 425–436.

Grovum, W.L. (1981), Factors affecting the voluntary intake of food by sheep. 3: The effect of intravenous infusions of gastrin, cholecystokinin and secretion on motility of the reticulo-rumen and intake, *Br. J. Nutr.*, **45**, 183–201.

Metz, J.H.M. (1975), *Time patterns of feeding and rumination in domestic cattle.* Veenman and Zonen, Wageningen.

Savory, C.J. (1979), Feeding behaviour, in *Food intake and regulation in poultry*, (Boorman, K.N. and Freeman, B.M., eds). British Poultry Science Ltd., Edinburgh, pp. 277–323.

Stephens, D.B. (1980), The effects of alimentary infusions of glucose amino acids or neutral fat on meal size in hungry pigs, *J. Physiol.*, **299**, 453–463.

4
DRINKING BEHAVIOUR

INTERNAL MECHANISMS

Thirst can be activated in three ways:

1. The renal system.
2. Changes in the plasma and vascular system.
3. Intracellular changes.

Renal output is matched to behavioural excesses or deficits so that taking in too much liquid is followed by the excretion of diluted urine and deficits in drinking are followed by the excretion of concentrated urine. Conservation is caused by the action of anti-diuretic hormone (ADH) from the posterior pituitary which acts directly on the kidney tubules to conserve water, and ADH output increases when blood volume decreases. Conversely, over-hydration of intravascular or intracellular fluids reduces ADH excretion. With small losses of blood, an enzyme, renin, is released from the cells in the walls of the afferent renal arterioles and this enzyme reacts with angiotensinogen to form angiotensin I which is hydrolysed into angiotensin II, a vasoconstrictor. This may elicit thirst by acting upon osmoreceptors or baroreceptors that monitor blood volume and which are situated in the right atrium, the aortic arch and in the carotid sinus. Volume regulation appears to be more important than osmoregulation, for during salt deficiency the body maintains its fluid volume and does not compensate for reduced salt levels by decreasing the volume. Large losses in volume are accompanied by the release of aldosterone from the adrenal cortex which acts to conserve salt and other minerals (Stricker, 1973).

Thirst can be caused also by cellular dehydration which results in compensatory drinking and which can be induced experimentally by the ingestion of hypertonic solutions of sodium chloride or sucrose. However, the amount drunk and the time spent drinking are not related in a simple way to the degree of cellular dehydration. In an experiment with dogs it was found that some over compensated when drinking after receiving hypertonic solutions, while others under compensated. However each individual was found to be very consistent. The cause of this variability between individuals is thought to be in renal mechanisms (Blass, 1973). In general, though, drinking to compensate for cellular dehydration is much more quickly accomplished than for vascular dehydration as two thirds of water ingested enters the cells and less than one tenth becomes plasma. Drinking usually stops before the hypovolaemic deficit is corrected (Stricker, *loc. cit.*).

The lateral pre-optic area of the hypothalamus is an osmoreceptive zone for thirst. Ablation of the area in rats and rabbits abolishes compensatory drinking in response to cellular dehydration and injections into it of NaCl in nearly physiological dosages produces drinking in hydrated rats. Conversely, injections of distilled water into the area will stop rats suffering from cellular dehydration from drinking, but not rats suffering from vascular dehydration. Andersson (1971) has shown that certain hypothalamic cells lining the third ventricle are sensitive to angiotensin. Previously, work in his laboratory showed that NaCl injections into that area in goats caused them not only to drink, but to perform a task that they had previously learnt in order to get water (Andersson and Wyrwicka, 1957). As stated earlier, hypovoleamia is detected by baroreceptors, and the impulses generated there are transmitted via the vagus and glossopharyngeal nerves to centres in the medulla oblongata and from there by ascending pathways that are involved in the regulation of blood pressure and ADH secretion.

There appears to be a very complex relationship between the cellular state, renal response and peripheral stimulation; rats that had been over hydrated while under anaesthetic then had their mouths washed with either water or hypertonic saline. Those that had undergone water irrigation urinated abundantly while those that had the saline treatment urinated significantly less (Blass, 1973).

The perception of water may vary according to the state of hydration or dehydration of the animal. Gentle (1974) found that chickens that had had 50 cm^3 of water introduced directly into the crop, when tested 15 minutes later, found the taste of water aversive.

Furthermore, oropharyngeal stimuli are important in maintaining drinking after destruction of the areas of the CNS that normally control it (Kissileff, 1973). The termination of drinking appears to be due to a synergistic action of oral and gut cues (Toates, 1979).

Not all drinking can be explained as functions of the types of dehydration described above. Eating is usually accompanied by drinking and this may be due to oropharyngeal sensations or due to the animal anticipating water deficits caused by changes in body fluids during digestion. Toates (*loc. cit.*) has suggested other factors that can lead to drinking and he includes, *inter alia*, the availability of water, habit and frustration. If water is difficult to obtain then the animal will reduce its intake, balancing intake and effort.

EXTERNAL STIMULI

In the case of farm livestock the releasers for drinking have only been investigated in the young chick. In a very old experiment Lloyd Morgan (1896) reported that naive chicks never peck at a sheet of water even if they are thirsty and standing in the water. However they will peck at any material or bubbles in the water and as soon as the bill is wet they will begin to drink. More precisely the releasers for drinking in chickens were investigated by Rheingold and Hess (1957). They assumed that any of the following properties of water, either alone or collectively, could attract the chick: colourlessness, transparency, reflective surface or movement. Six stimuli were chosen on the basis of possessing some of these properties.

A total of 100 three day old chicks were used. The experimental birds were deprived of food until after their first test and were tested again at 7 days of age. Twenty eight control chicks which had had experience of food and water were also used. Each bird was tested individually for its initial preference from the following stimuli: water, blue water, red water, polished aluminium, mercury and colourless plastic. Their order of choice at 3 days was mercury, plastic, blue water, water, aluminium and red water. Three times as many birds chose mercury in preference to water and more than twice as many chose plastic. At 7 days of age the order was only slightly changed with blue water superceding plastic and red water supplanting the metal. The control chicks were found to have the same preferences. Mercury in this case is what is known as a super normal stimulus. Other examples from other motivational systems are known, e.g. the oyster catcher will try to retrieve an egg almost as large as itself in preference to its own egg provided the artificial egg is coloured the same as the normal egg.

However, drinking behaviour is not only released by visual cues. Thirsty rats will lap at a jet of cool air (Kissileff, *loc. cit.*). In this case the stimulus is a thermal one.

CONSUMMATORY BEHAVIOUR

As in feeding behaviour the amount of consummatory behaviour is not necessarily related to the degree of thirst. As in the case of feeding behaviour consummatory feedback mechanisms operate once liquid enters the mouth. Miller (1957) found that water injected into the stomach led to a decrease in thirst, as measured by either rate of bar pressing or by the amount of water drunk. Water taken in by the mouth however produced an even greater decrement in the willingness to work for water.

SOME CONSIDERATIONS IN RELATION TO FARM LIVESTOCK

Rowland (1980) has suggested that non-physiological factors may govern drinking in the laboratory rat but not the wild rat, which may drink only when physiologically motivated to do so. Intensively and semi-intensively housed farm animals may resemble the laboratory rat in this respect, while stock kept extensively with distant sources of water may respond mainly to physiological stimuli. Under grazing conditions the frequency of drinking will depend on temperature, condition of the food and the distribution of water as well as type of breed (Arnold and Dudzinski, 1978). Under intensive and semi-intensive conditions social factors will play a large rôle, although drinking is usually associated with feeding. Animals will usually drink shortly after eating, but if thwarted from feeding they often drink instead and we shall discuss this in later chapters. However the composition and texture of the diet as well as ambient temperatures will also affect the pattern and frequency of drinking.

POINTS FOR DISCUSSION

1. Would the feeding of a diet in a pelleted form instead of a mash

form alter the drinking behaviour of animals? If so, how would it cause such a difference?

2. Discuss the physiological and behavioural mechanisms that cattle and sheep might employ to conserve water in hot dry areas.

FURTHER READING

Arnold, G.W. and Dudzinski, M.L. (1978), *Ethology of free-ranging domestic animals*. Elsevier, Amsterdam, pp. 43–48.

Winchester, C.F. and Morris, M.J. (1956), Water intake rates of cattle, *J. Anim. Sci.*, **15**, 722–740.

Ingram, D.L. and Stephens, D.R. (1979), The relative importance of thermal, osmotic and hypovolaemic factors in the control of drinking in the pig, *J. Physiol.*, **293**, 501–512.

5
SEXUAL BEHAVIOUR

INTERNAL MECHANISMS

The sexual behaviour of the male usually consists of courtship followed by copulation. In order to analyse his sexual behaviour it is necessary to keep this dichotomy in mind and also to divide copulation by the male into its components of mounting, intromission and, depending on the species, possibly thrusting, and, finally, ejaculation. In the case of the female a form of courtship may exist but be difficult to discern. Generally oestrous females become restless; the bitch on heat, for example, is well known to run loose if allowed to. This active participation of the female in sexual behaviour is defined as proceptivity and two other aspects of her role can also be distinguished: her receptivity and her attractiveness to males (Beach, 1976; Keverne, 1976.) Such concepts have not been applied as yet to the females of our common livestock species, but they might be useful if we are to understand their behaviour more fully. In most agricultural species successful copulation depends on the female adopting a copulatory stance in order for the male to gain intromission, and this stance as we shall see is brought about by a coordinated series of events.

For a long time testosterone was considered to be the hormone governing male sexual behaviour. Early work showed that injections of testosterone into male chicks resulted in the males showing mounting, treading and cloacal contact as well as crowing (Hamilton, 1938). Likewise Craig *et al.* (1954) increased the sexual performance of some male rats by injections of testosterone. However, more recently Fletcher and Short (1974) have shown that male mating behaviour can be induced by oestrogen injections and so possibly it is not testosterone itself that is responsible for male mating behaviour, but androsterone or an oestrogen derivative for, as shown in Fig. 5.1,

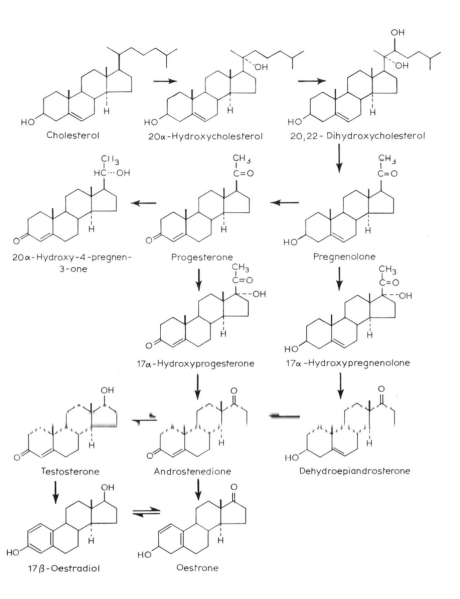

Fig. 5.1 The biosynthetic pathway for gonadal steroid hormones. (From Hunter, 1980.)

testosterone can be aromaticized to oestradiol. Castration of the male does not always lead to a cessation of sexual behaviour. Tom cats that have had sexual experience before castration continue to show copulatory behaviour, while those that have had little sexual experience before the operation show a rapid or total decline after it (Rosenblatt and Aronson, 1958). This persistence in sexual experience is considered to be due to CNS factors rather than to steroids produced by the adrenal glands. Castration of the female, however, results in an immediate loss of sexual behaviour in all cases.

Modern methods of assaying hormones has shown that the relationship between the amount of testosterone circulating in the blood plasma and level of sexual behaviour is not a simple one. Differences between rams have been found in the levels of circulating testosterone and within an individual these levels may vary from hour to hour (Fig. 5.2 and Katongole *et al.* 1974). Such episodic release of

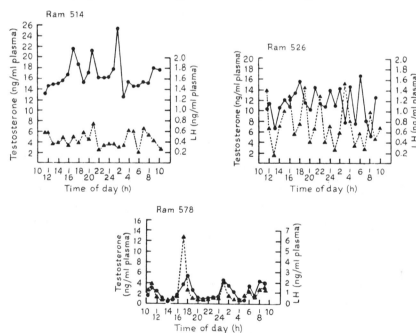

Fig. 5.2 The concentrations of luteinizing hormone (LH) (▲) and testosterone (●) in the peripheral blood of three rams, measured at hourly intervals on 30–31 October (rams 514, 526) and on 30–31 May (ram 578). (Katongole *et al.*, 1974.)

testosterone has also been reported in mice by Batty (1978) who also found that certain strains differ in the mean amount of testosterone circulating in the plasma of the males. Furthermore the mean male sexual behaviour scores of the strains differed significantly, and were negatively correlated to the levels of plasma testosterone. She offers several explanations for this: the strains with the low levels may in fact have a high rate of release of the hormone from the testes as well as a rapid clearance rate, or the areas of their brains mediating sexual behaviour might have heightened sensitivity to the hormone. Various rates of aromatization may also be involved. Such differences may account for individual differences, although social factors are also likely to be implicated.

Pre-optic and hypothalamic areas appear to be involved in the regulation of male sexual behaviour in a wide variety of vertebrate species (Kelley and Pfaff, 1978). Evidence from the male Rhesus monkey suggests that the pre-optic area may be involved in the processing of sexual stimuli, for males with lesions in this area masturbate and ejaculate frequently but have no interest in the oestrus female.

Oestrogen and progesterone injections can bring spayed guinea pig females as well as ewes into oestrus (Grunt and Young, 1952; McGill, 1978). A study on cows showed that the oestradiol concentration increases during the three days preceding oestrus and attains peak values (17 ± 1.9 ng per 100 ml plasma) four hours before oestrus can be detected. On the day of oestrus the level declines to reach a nadir (0.8 ± 0.11 ng per 100 ml) before the time of ovulation. A minor peak is found on day 4 of the 21 day cycle and a greater peak which occurs on day 11 (8.1 ± 3.6 ng per 100 ml) may precede the pre-ovulatory surge of luteinizing hormone (Fig. 5.3 and Shemesh *et al.*, 1972). In the female monkey both oestradiol and testosterone control the receptivity of the female but their actions are somewhat different. While both probably act on the CNS, the oestradiol also has a direct lubricating function on the vagina. Together they enhance proceptivity while progesterone suppresses it. The androgens are most probably derived from the female's adrenal glands (Keverne, 1976).

The hypothalamus is also important in mediating sexual behaviour in the female. For example, Harris and Michael (1964) were able to induce oestrus in female cats by implanting 0.1 mg of stilboestrol (a synthetic oestrogen) into the posterior hypothalamus whereas a dose

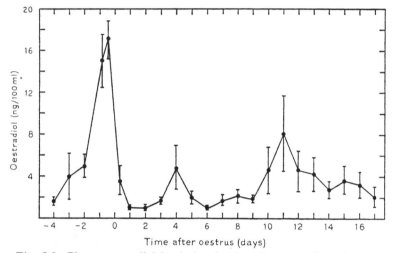

Fig. 5.3 Plasma oestradiol levels in nine lactating cows throughout the oestrous cycle. The vertical lines indicate ± SEM. (From Shemesh *et al.*, 1972.)

ten times as large had no effect when given subcutaneously. However, their genital tracts showed none of the changes that normally accompany oestrus, although these females displayed oestrus behaviour. In general, areas of the brain involved in the sexual behaviour of mammalian females appear to differ somewhat between species (Kelley and Pfaff, 1978). In rats these are the medial pre-optic area, and anterior hypothalamus; in hamsters, the hypothalamus and ventromedial nucleus. The latter areas are also involved in the female guinea pig together with a medial hypothalamus area and the lateral arcuate, while in the rabbit the pre-mammillary and mammillary bodies are also involved.

The generalized system that emerges for females is the following (Bastock, 1967):

Environmental factors (seasonal or stimuli from the mate) →
Ovarian or ovulatory centre →
Neurohormones (e.g. LH and FSH releasing factors) →
Pituitary ((a) FSH → Oestrogen (b) LH → ovulation).
Oestrogen → Behaviour centres.

In young animals this chain of events may be incomplete. For instance, Edney *et al.* (1978) reported that 4 out of 61 lambs ewes failed to ovulate at first oestrus. In some species, such as the rat, copulation may then lead to further LH secretion, (Linkie and Niswender, 1972), whereas in others, such as the rabbit, ovulation will only take place if copulation occurs. In the domestic hen there appears to be no correlation between rate of copulation and rate of ovulation (Wood-Gush, 1958a). The notion of a behaviour centre does not mean that other parts of the brain are not involved, once the centre has been activated. For instance, the forebrain is necessary to obtain fully integrated mating behaviour in the female cat. Females whose forebrain has been severed from the mid-brain will stand for copulation on stimulation of the vulva, but their behaviour has not the spontaneity of the normal female.

EXTERNAL FACTORS

In seasonally breeding species, once these endocrine mechanisms have begun operating, appetitive behaviour will begin. Males will begin to seek females. This may mean, in the case of the Soay rams on St Kilda, a ram leaving the male groups, or in the case of the Bighorn sheep, a male walking many miles to his individual mating area. The females too will show appetitive behaviour; cows become more active and ewes may seek out rams (Lindsay and Fletcher, 1972); they show proceptivity although they do not appear to increase their general activity.

The early ethological literature emphasized that the species under study responded to very specific stimuli, the male stickleback to the swollen belly and posture of the fertile female, rather than to her whole configuration (Tinbergen, 1951). Studies with farm animals continue to bear this out. Smell, for example, may be used by the bull to locate oestrus cows (Craig, 1980) while the releaser for mounting behaviour in the bull is an inverted U about the size of a cow's rump (Hale, 1966). Through conditioning other stimuli become important; experienced bulls when blindfolded can be induced to mount by tactile stimulation of the chest (Hale, 1966). Lindsay and Fletcher (1968) investigated the sensory modalities employed by the ram in seeking oestrus ewes. Rams that could hear, smell and see the ewes had a success rate of 95%; deaf rams a success rate of 86%, but in animals deprived of their sense of smell and in blindfolded ones the rate fell to 73% and 49% respectively. This type of experiment can only give an indication of the involvement of the different stimuli from

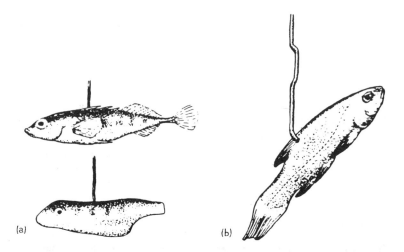

Fig. 5.4   (a) Two models of female three-spined stickleback. Detailed model with 'neutral' abdomen (above); crude model with swollen abdomen (below). (b) Dead tench (*Tinca vulgaris*) of stickleback size presented in attitude of readiness of female three-spined stickleback. Only the latter two models release courtship behaviour in the male. (After Ter Pelkwijk and Tinbergen, 1937; Tinbergen 1951.)

the female, for the procedures used might upset the motivation of the males. In the normal course of events it is possible that different stimuli may come into play at different times. One may enable him to locate the female, another may guide him to orientate himself for courtship, yet another to mount and yet others to complete the chain of behaviour. The stimuli that release sexual behaviour in the turkey tom and domestic cock were compared by Carbaugh *et al.* (1962). The head of the female is the primary stimulus for sexual arousal and orientation for mounting in the turkey tom while in the cock the head and body of the female have equal stimulus value for sexual arousal and orientation. The complete copulatory pattern is released mainly by the female head in the turkey and by the body in the case of the rooster (see Table 5.1).

Special odours or pheromones may act as releasers or key stimuli in the mating behaviour of many species. In the case of insect pests, they are used as lures by man. The role of vaginal secretions as pheromones in the sexual behaviour of the Rhesus female monkey has been

Table 5.1 Relative stimulus functions of the head, body and tail on turkeys and chickens. Maximum value $= ++++$. (From Carbaugh *et al.*, 1962.)

		Turkey	*Chicken*
Head	Arousal	$++++$	$+++$
	Complete pattern	$++++$	$+$
Body	Arousal	$+++$	$+++$
	Complete pattern	$+$	$+++$
Tail	Arousal	No	$+$
	Complete pattern	data	0

investigated in detail. The secretions consist of volatile fatty acids: acetic, propionic, isobutyric, butyric and isovaleric acids, and it has been possible to make a synthetic pheromone to test males. The effects of this pheromone on the males is complex; not all males react in a similar manner and their reaction varies with different females. It is concluded that odour cues are unnecessary for copulation to occur but may assist the male in determining the sexual status of his females (Keverne, *loc. cit.*).

In farm livestock the most complete analysis of the stimuli from the male that lead to the copulatory stance by the female is in the case of the gilt (Signoret 1971). When pressure on her back is supplied by man 49% will respond by standing for mating. If this pressure is accompanied by recordings of the boar's courting sounds then 71% will give the response and if pressure from a man is accompanied by the scent of the boar 81% will respond. If these tactile auditory and olfactory stimuli are combined 90% of gilts show the mating stance. If a fourth stimulus, the sight of a boar is included 97% respond. It is unlikely that the female is responding to the total form of any of these types of stimuli and most probably is responding, for example, to only part of the male's vocalizations and scent. This type of interaction between stimuli which have an additive effect is known as the heterogeneous summation of stimuli systems and has been found in other species, and was mentioned in Chapter 3.

Table 5.2   Effectiveness of various stimuli in releasing the postural standing reaction of the oestrous sow. (From Signoret *et al.*, 1975.)

Stimulation	Postural standing reaction (%)
Pressure on back only	48
Broadcasting boar's courting grunts	70
Odour of preputial secretion at 40° C	80
Smell and sound of boar	90
Smell, sound and sight of boar	97

COURTSHIP AND COPULATION

Most birds and mammals show courtship behaviour and the behaviour shown may, as in the case of the gilt, involve at least four sensory modalities. In that example it can be seen that the male's courtship serves to bring the female to the state of readiness to copulate. In addition, in some species the female actively solicits the male and in the case of the ewe she may also nuzzle his flank or scrotum (Edney *et al.* 1978). Analyses of courtship behaviour of birds and fish species that have been extensively studied indicate that courtship displays, like the social displays discussed earlier, greatly resemble the behaviour of animals in conflict between two or more incompatible moods (Bastock 1967). In many cases the conflict is between aggressive and sexual reactions and, in some, fear may be a third conflicting motivation. This analysis of courtship behaviour will be dealt with more fully when we come to discuss the evolution of behaviour, but at this point it is worth emphasizing that copulation is usually the end-point of an elaborate chain of behavioural sequences, that serve to overcome aggression and fear which are usually elicited by the very close proximity of another animal. Indeed courtship is most marked, for example, in the domestic cock when he is placed with strange hens (Wood-Gush, 1956). In general the courtship of our poultry species follows the pattern of many other birds in which the origins of the display can be discerned, but in the case of our other common species of livestock it is not so obvious. Nevertheless the theme may be discerned in the very early phases of courtship when the animals are approaching one another. Their later and more obvious courtship behaviour patterns which serve to stimulate the reproductive organs of the pair may have a different evolutionary derivation.

Like appetitive behaviour in other motivational systems, courtship can be altered by experience. An interesting example of adjustment of courtship through learning was found in the Burmese Red Jungle fowl (Kruijt, personal communication). This bird, like some domestic fowls, performs a behavioural pattern in courtship called 'cornering' in which the male in a pen goes to a corner, lowers himself with exaggerated scraping movements of the feet and emits a particular series of short staccato calls. It is usually performed relatively infrequently and usually late in the male's repertoire. However Kruijt was able to alter its occurrence by allowing the male access to a hen only if he performed it. A female was placed in a pen under a basket where she could be seen and the male released into the pen. At first the male courted the female using the behavioural patterns common in courtship but eventually he 'cornered' and the female was immediately released from the basket. After a very few such trials the male 'cornered' immediately on entering the pen, thus indicating the relative flexibility of courtship. Even with such flexibility it is unlikely that courtship is purely appetitive for it appears to be reinforcing in its own right, and not merely through its close connection with copulation. Indeed some animals will learn to perform operant tasks in order to court a female of their species.

As mentioned earlier a behavioural analysis of copulation should consider the various components of the behaviour; in the case of the males, orientation, mounting, thrusting and ejaculation. The sequence can stop at any of these points. Studies on the various components of copulation seem to suggest that it is not a single unitary motivational system. Beach and Jordan (1956) postulated that, in the case of the male rat, there are two mechanisms at work; one concerned with the arousal of sexual behaviour (the Arousal Mechanism) and the other with the execution of the copulatory pattern (the Copulatory Mechanism). Similarly, as mentioned earlier, Batty (*loc. cit.*) in her study on the relationship between plasma testosterone levels and sexual behaviour in male mice made a similar division. One set of measurements taken by her was similar to the Arousal Mechanism postulated by Beach and measured the initiation of the sexual response, while the other involved ejaculation and corresponded to the Copulatory Mechanism. During the course of the experiments, it was found that the Arousal Mechanism was more closely correlated, albeit negatively, with the testosterone plasma levels than the Copulatory Mechanism, thus suggesting somewhat different controlling mechanisms. More recently Sachs and Barfield (1976) reviewed the work on the sexual behaviour of the male rat and concluded that probably more than two mechanisms control it. Although this concept deals with rodents it may have a wider application.

GENETIC FACTORS CONTROLLING MATING BEHAVIOUR

Genetic studies of behaviour patterns and particularly of sexual behaviour provide a method of analysing the relationships between the various components of what might be otherwise thought of as a unitary system, for they may provide a 'natural dissection' of the behaviour. Goy and Jakway (1959) and Jakway (1959) investigated the genetic basis of variation in the mating behaviour of guinea-pigs (Fig. 5.5). Two parental strains named strains 2 and 13 which differed in certain respects were used. The females used in the tests were all spayed and hormonally treated with standardized doses. However strain 2 females were more responsive to the treatment than the strain 13 females: they came into oestrus sooner but the strain 13 females once in oestrus were more vigorous in their responses. Experiments involving strain-crosses and back-crosses revealed that latency to oestrus, duration of oestrus and the percentage of females responding were all correlated genetically. However the difference between a long and a short period of maximum receptivity depends on a single gene; the short duration being determined by the recessive allele. Male-like mounting on the other hand is determined by a large number of genes. In the case of the males, strain 2 males mounted at significantly higher rates than the strain 13 males and the crosses revealed that the lower mounting rate was dominant over the higher mounting rate, i.e. the strain 13 genes were dominant for this trait. In the course of copulation the male guinea-pig has several intromissions before ejaculating and the two strains differed both with respect to intromission rate and rate of ejaculation under standard tests. The genetic analysis showed that both higher intromission rate and higher ejaculation rate were the dominant traits. Such differences in genetic determinations as stated earlier indicate that sexual behaviour in this species at least is not a unitary system.

In the domestic fowl two similar genetic studies have been carried

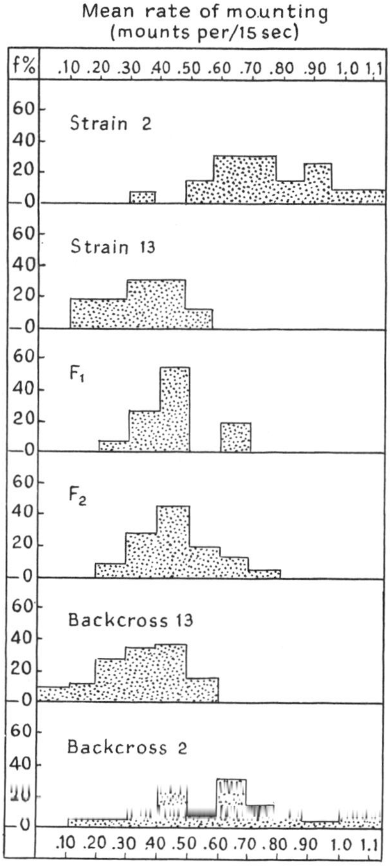

Mean rate of mounting
(mounts per/15 sec)

Fig. 5.5 The distribution of frequencies of mounting in the males from the two parental strains and the crosses between these two strains. The genes from the strain with the lower rate of mounting appear to be dominant. (From Jakway, 1969.)

out. Wood-Gush (1960) selected two lines of Brown Leghorn cockerels, one selected for a high rate of copulation and another for a low rate of copulation under identical conditions. Initially they had no differences in semen production but in the F_2 generation the low line males were producing a mean of 0.455 ± 0.043 cm^3 per ejaculate compared with a mean of 0.16 ± 0.046 cm^3 per ejaculate by the high line males. Their copulation rates per test at this time were completely opposite: 11.7 ± 0.79 for the high line and 6.25 ± 0.74 for the low line. The males of both lines mated immediately on being put into the test pens, suggesting that there were no differences in their Arousal Mechanisms. However when copulation *and* ejaculation were investigated together, it was found that whereas the low line males ejaculated at every copulation, the high line males did not always ejaculate. This begs the question as to which strain was really highly sexually motivated. It can be argued that the low strain males were the most highly aroused as they reached ejaculation sooner, whereas the high strain males had to mount, tread and have cloacal contact with several females before they were sufficiently aroused to ejaculate. Furthermore, if one assumes a dual controlling mechanism with an Arousal Mechanism and a Copulatory Mechanism then the differences might lie in the Copulatory Mechanism which would be slower to act in the high strain males than in the low strain males. Siegal (1972) carried out a very similar selection experiment using White Leghorns on a much larger scale over 11 generations. Like Wood-Gush he obtained clearcut responses to selection for high and low mating activity; semen production and mating activity also diverged similarly. He scored the cumulative number of completed matings (i.e. those that ended in cloacal contact) in eight ten-minute periods as one of his selection criteria. Other measures were mounts and treadings, the latter being copulatory acts in which the cock mounts and treads the hen and then dismounts without completing the copulation. Analysis showed that while there were some genes common to all these components, there were others that were unique to each component, thus indicating that selection on a commercial basis for sexual behaviour will not yield satisfactory results.

EARLY EXPERIENCE AND THE DEVELOPMENT OF SEXUAL BEHAVIOUR

Brief mention was made earlier of sexual imprinting. Originally it was considered that sexual imprinting coincided with filial imprinting but evidence now suggests that it takes place at a later stage. Vidal (1980) kept individual cockerels isolated from hatching and divided them into 3 groups which underwent different training regimes with a parental model. Group 1 was exposed to the model singly from the day of hatching to day 15. Groups 2 and 3 had similar training on days 16–30 and 31–45 respectively. During training group 3 cockerels showed

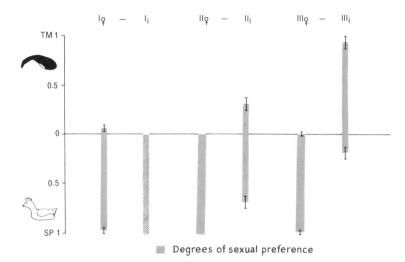

Fig. 5.6 TM = Training model. SP = Stuffed pullet. Groups I, II and III had experience with TM from days 1–15, 16–30 and 31–45 respectively. The symbol ♀ signifies that the males from those groups were each kept with a female of similar age. The symbol i indicates that these males were reared in isolation. All males were tested during days 150–165. (From Vidal, 1980.)

little interest in the model, but when tested as young adults at 5 months of age and given a simultaneous choice between the training model or a stuffed pullet or another model, group 3 males courted the training model, whereas the males of the other groups showed little sexual behaviour to the training model.

Immelmann (1980) cites a number of examples. In one experiment on sexual imprinting (Sonneman and Sjolander, 1977), zebra finch females were raised with Bengalese finch parents and after the sensitive period they were kept individually isolated. When sexually mature, they were tested to ascertain their preference between the males of their foster species and of their own species. Only males that were imprinted on zebra finches were used. These fostered females showed a preference for Bengalese finch males although they did not prefer them exclusively to zebra finch males. Such preferences seen in imprinted animals raise the question about the similarity of the stimuli that attract them to the foster species and the homologous stimuli in the parental species. Räber (1948) described the behaviour of a turkey tom that was imprinted on humans and tended to attack ladies with hanging shoulder bags, which Räber suggests were equated by the bird to the hanging appendages of the turkey tom. This anecdotal account suggests that from our point of view wide generalization must occur. However, if the animals are responding to only a very limited part of the stimulus then the amount of generalization might not be great. Preferences shown by domestic hens were studied by Lill and Wood-Gush (1965), using hens of different breeds that had plumage differences as well as size differences. Except for the females of a variegated breed with no typical colour, the females preferred males of their own breed. In a later experiment Lill (1968) showed that if females were reared from hatching with birds of the other breed then there was no discrimination. Preferences of this sort are found in other species as well (Craig, 1980).

Before the onset of sexual imprinting in fowls, chicks are able to perform the male copulatory motor patterns. Andrew (1966) reported juvenile copulation by two male chicks aged forty-eight hours, when the experimenter's hand with fingers extended, was introduced into the chick's cage and thrust towards the bird (Fig. 5.7). A number performed partial copulatory behaviour and at a slightly older age a small number of female chicks assumed the male role in the experimental situation. Testosterone injections were unnecessary to elicit this behaviour and even appeared to suppress it when injected into female chicks.

Juvenile tidbitting was also observed. Hence the neural mechanisms necessary for the performance of copulatory behaviour are present from an early age, but normally full copulation would not be seen because of lack of cooperation on the part of the other chicks. One performance of complete juvenile copulation delays further copulation for some time as with an adult, but the feedback mechanisms are not necessarily the same, since the young birds cannot ejaculate. Possibly constriction of the vas deferens, which occurs, may act as a reinforcer (Andrew, *loc. cit.*). However such changes in the vas deferens would not account for females, adults or chicks, performing the male role. Andrew suggests that tactile stimulation from the ventral surface acts as a reinforcement for future mountings.

Andrew's chicks had been mainly kept in isolation from the time of hatching or from a very early age, and it is unlikely that juvenile copulation occurs in birds kept in groups, although attempted

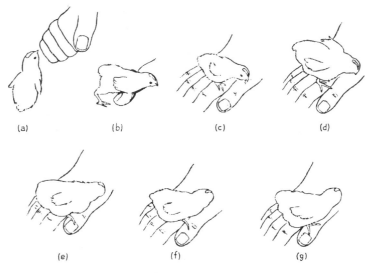

Fig. 5.7 Juvenile copulation. (a) Erect, ready to leap up at hand, which is approaching. (b) Climbing on to hand immediately after it is lowered. Note chest is in contact with hand and bill is lowered to grasp at hand. (c) Bill grasps hand. (d) Wings are raised slightly, as pelvis lowering begins. (e)–(g) Treading, which caused the shift of foot position from (d) to (e), passes into quivering movements of the body. (From Andrew, 1966.)

copulation may be common. The development of normal sexual behaviour in heterosexual or unisexual groups has not been studied as fully as it deserves. The best observations are those of Kruijt (1964) on the Burmese Red Jungle fowl and, although not identical to domestic fowl, their behaviour may act as a guide. Between the ages of thirty and eighty days copulatory behaviour may occur unmixed with aggression and escape, but no complete copulation was seen in this age group. Certain copulatory patterns are mainly performed by low ranking immature males, and Kruijt suggests that this is because, in these males, escape and aggression are more finely balanced than in high ranking males, in which they would be masked by aggression. From 80 to 120 days the infantile calls change and adult calls are heard. Complete copulation does not occur, and copulatory behaviour does not change much in form, although more of it goes through to the treading stage. Mounting is now released by both standing as well as sitting birds of either sex and is no longer directed exclusively to birds of lower rank. Most of the mounting was found by Kruijt to be done by lower ranking males.

From 120 days the males start to copulate successfully and this is often mixed with strong tendencies to attack, for kicking of the female during copulation was often witnessed by Kruijt. Escape behaviour may also be present and tidbitting and 'cornering' appear in the cockerel's repertoire. From a detailed analysis of temporal patterning of displays and examination of their components, Kruijt concludes that the best working hypothesis to explain the causation of the main displays is that sexual, attack, and flight systems are continuously activated during encounters between a male and a female during this stage of the bird's life.

In the young male piglet, components of its sexual behaviour appear at different ages (Ruth Hutton, personal communication). For example ano-genital sniffing may be seen in 12 day old piglets but it only becomes firmly associated with sexual behaviour at 3–5 months. By the latter age males may approach sows with courting grunts and sniff the sow's ano-genital area. Also at this age both male and female juveniles may sniff and lick the urine from females. Mounting has been seen in piglets as early as 7 days of age but the orientation is incorrect. The earliest mounting with correct orientation was seen at 21 days of age. Pelvic thrusting was first observed at about 3 months of age and first occurs with mounting about one month later. It is common for pairs of 6 month old males to perform long bouts of mounting one another. Similarly at this age they also mount young females and pay a great deal of attention to sows in oestrus.

Great variability in the frequency at which young males show sexual behaviour is seen in piglets and chicks. Wood-Gush (1963) treated young male chicks with testosterone and scored their sexual behaviour in controlled conditions and again later when they were sexually mature, but failed to find any correlation between their scores at the two ages.

Early social experience can also affect mating behaviour in animals. Hemsworth and Beilharz (1979) reared three groups of 6 boars each from 20 days to 7 months of age under one of the following social conditions: (a) a mixed sex group, (b) an all male group and (c) in social restriction with no visual or physical contact with other pigs. When 7 to 13 months old their sexual behaviour was studied. The total number of copulations and the total number of all the courtship behaviour activities were significantly greater for the boars reared in

Fig. 5.8 Elements of copulatory and aggressive behaviour being performed by young Burmese Red Jungle fowl males, aged between thirty and eighty days towards sitting chicks. (a), (b) and (c) copulatory; (d) copulatory attack; (e) and (f) attack; (g) and (h) exploratory. (From Kruijt, 1964.)

groups 1 and 2 than for those reared in social isolation. Growth rate to 30 weeks of age, testicle size and semen quality were unaffected by treatment. The crucial factors that led to these differences were not identified and there is need for further research in this area.

SOCIAL FACTORS AFFECTING SEXUAL BEHAVIOUR

Small scale experiments with domestic fowls (Guhl *et al.*, 1945; Guhl and Warren, 1946) indicated that cocks higher in the social order performed most of the matings. Indeed in the first study the top ranking male's dominance was so complete that when he was removed, the remaining cocks, which had mated frequently before being grouped, then failed to mate at all; they appeared to be 'psychologically castrated'. Recently Kratzer and Craig (1980) carried out an experiment of this type, investigating the effect of social rank on mating behaviour in males in flocks of different sizes and kept at different densities. Over the first 5 weeks of the experiment, the dominant males performed most of the completed matings and were less interrupted while mating than the other males. Nevertheless the other cockerels, comprising 67% of the males in the experiment, executed 58% of the completed matings. In sheep, dominance is also important when the males are in restricted areas such as pens, but under extensive conditions it has little effect. With bulls, however, dominance is effective under both intensive and extensive conditions (Craig, 1980). In the domestic cock and the bull in which dominance is important in relation to the number of offspring sired, it might be expected that there would be increased aggression over the generations. However this is plainly not so and several explanations may be found for this. Males that are too aggressive will frighten the females and thus leave less offspring than expected. Also, aggression is not synonymous with dominance. In fowls body size and fighting skill are important (Wood-Gush, 1971) and in bulls age as well as body size and length of residence in the herd are likely to be important factors in determining dominance.

SOME CONCLUSIONS IN RELATION TO FARM ANIMALS

We have seen that the concepts of releasers and key stimuli operate in the mating behaviour of the species that have been investigated to an appreciable extent. It is possible, therefore, that the failure to conceive on the part of some female farm livestock may be due to the fact that they are not provided with adequate stimuli for us to gauge their reproductive state correctly, in cases in which artificial insemination is used. Likewise some males may not be sufficiently stimulated. Natural matings may fail because of deprivation of social experience in early life or because of lack of stimulation from the opposite sex, as we saw in the case of the young boars. The dominance hierarchy too may affect the performance of males under competition so that the number of males and females in a given area may be important. Preferential mating may also affect fecundity.

POINTS FOR DISCUSSION

1. Discuss the factors that could lead to preferential mating in a species of farm livestock.
2. Mention some factors that might lead to a low rate of mating in the bull, boar and ram under various types of husbandry.
3. Describe the role of courtship behaviour in mating and describe it in one species of farm livestock.
4. Discuss the genetic control of mating behaviour and try to assess its practical use in relation to one species of farm livestock.

FURTHER READING

Banks, E.M. (1965), Some aspects of sexual behaviour in domestic sheep, *Ovis aries, Behaviour*, 23, 249–279.

Beach, F.A. (1968), Coital behaviour in dogs. III. Effects of early isolation on mating in males, *Behaviour*, 30, 218–238.

Beach, F.A. (1970), Coital behaviour in dogs. VIII: Social affinity, dominance and sexual preference in the bitch, *Behaviour*, 36, 131–148.

Hale, E.B. (1966), Visual stimuli and reproductive behaviour in bulls, *J. Anim. Sci.*, 25 (suppl.), 36–44.

Hemsworth, P.H., Beilharz, R.G. and Galloway, D.B. (1977), The influence of social conditions during rearing on the sexual behaviour of the domestic boar, *Anim. Prod.*, 24, 245–251.

Hemsworth, P.H., Winfield, C.G., Beilharz, R.G. and Galloway, D.G. (1977), Influence of social conditions post-puberty on the sexual behaviour of the domestic pig, *Anim. Prod.*, 25, 305–309.

Hemsworth, P.H., Findlay, J.K. and Beilharz, R.G. (1978), The importance of physical contact with other pigs during rearing on the sexual behaviour of the male domestic pig, *Anim. Prod.*, 27, 201–207.

Hemsworth, P.H. and Beilharz, R.G. (1979), The influence of restricted physical contact with pigs during rearing on the sexual behaviour of the male domestic pig, *Anim. Prod.*, 29, 311–314.

Key, C. and MacIver, R.M. (1980), The effects of maternal influences on sheep: breed differences in grazing, resting and courtship behaviour, *Appl. Anim. Ethol.*, **6**, 33–48.

Lill, A. and Wood-Gush, D.G.M. (1965), Potential ethological isolating mechanisms and assortative mating in the domestic fowl, *Behaviour*, **25**, 16–44.

Wood-Gush, D.G.M. (1954), The courtship of the Brown Leghorn Cock, *Brit. J. Anim. Behav.*, **2**, 95–102.

6
PARENTAL BEHAVIOUR

In this chapter not only post-natal behaviour but pre-parturient behaviour will be considered, and we shall consider in some detail four species: two avian and two mammalian. While these do not cover a wide cross section of all types of parental behaviour, they do show some of the variation found in the two classes of animals in this behaviour and its controlling mechanisms. Two are common agricultural animals, the domestic hen and the ewe. Further examples illustrating various facets of parental behaviour in farm animals are listed at the end of the chapter.

THE DOMESTIC HEN

The physiological basis of nesting behaviour

The hen lays her eggs in batches or clutches with an interval of about 24–28 hours between each egg. A clutch may be followed by one or more pause days on which no eggs are laid. Sometimes the interval between eggs may be less than 24 hours, but generally each egg in a clutch is laid somewhat later each day. In hens with high productivity the clutch may consist of 30 eggs or more and the interval between eggs may be 24 hours plus or minus a few minutes.

Egg laying behaviour starts usually 1–2 hours before oviposition with the hen showing some restlessness, which in a pen with trap-nests may mean that she will pace around the pen peering up at the walls, calling frequently. Similar calls are often given by pullets in the days before their first egg and the calls can be induced in young immature females by oestradiol monobenzoate. The restlessness of the adult female about to lay is absent in these young treated females, although they orientate themselves away from their pen mates while calling (Wood-Gush and Gilbert, 1969a). Periods of restlessness are interspersed with periods of feeding and drinking but eventually the hen begins to take an interest in the trap-nests, peering into them, and eventually enters one and sits. Some time after oviposition she usually cackles and stands waiting to be released from the trap-nest. The nest entered by a hen is usually the same one each time; great conservatism is shown, so that the same nest may be used by a hen in successive years (Wood-Gush, 1954). Nest 'examination', nest-entry and sitting are under the control of the post-ovulatory follicle (Wood-Gush and Gilbert, 1964). With many hens the temporal pattern of egg laying within a clutch is very regular and ovulation of the next egg can be predicted. If the follicle from which that egg is derived is excised or ligated after ovulation, the normal behaviour patterns do not occur in relation to the laying of that egg. The egg is usually dropped by the hen about 48 hours later without any nesting behaviour. Table 6.1 shows that various other treatments involving similar surgery do not affect the behaviour patterns. It is assumed that oestrogen and progesterone are the agents responsible for the behaviour, as they will produce the behaviour in ovariectomized hens injected with those hormones. A striking feature of this controlling mechanism is that the physiological stimulus (ovulation) occurs about 24 hours before the behaviour, and the behaviour will occur regardless of the fate of the

Table 6.1 The effect of removing or ligating the post-ovulatory follicle on the nesting behaviour of the hen compared with other (control) treatments. The operations were performed about 3 hours after ovulation and about 22 hours before the egg was due to be laid. (From Wood-Gush and Gilbert, 1964.)

Treatment	Number of hens with nesting behaviour	
	Affected	*Unaffected*
Main follicle removed	12	5
Main follicle ligated	13	4
Main follicle manipulated	0	20
Penultimate follicle removed	3	22
Immature follicles and part of ovarian wall removed	2	15

egg. If the egg is extruded prematurely, either through artificial or natural means, the bird still goes through the behaviour at the expected time. In battery cages many birds show the initial restlessness and calling and tend to show the conservatism found in nest selection by laying in the same corner of the cage (Wood-Gush and Gilbert, 1969b). However the restlessness may give way to escape behaviour or to vacuum nest-building; this will be discussed later when we deal with conflict behaviour and with animal welfare.

Nest-building and the nest site selection

Hens living in the wild were also found to show some pre-laying locomotion and calling before going off to well-concealed nest sites (Duncan *et al.*, 1978). Nest site selection and nest-building were studied indoors by Wood-Gush (1975a). He allowed hens access singly to a pen with litter and feathers on the floor several hours before they were due to lay. After examining the pen with much locomotion and calling, each hen chose a site and began to make a nest in the litter by rotating in the litter, pushing the feet outwards. Then lowering herself so that her keel touched the litter, she would rotate several times until a shallow bowl-shaped nest appeared. Then she would leave the nest, pick up feathers and bits of litter throw them onto her back and, carrying material in her bill, return to the nest where with subsequent rotation movements the material would fall from her back onto the top of the nest. Gathering was also done by the bird sitting and reaching out for nest material which was thrown onto her back or raked towards her with her bill. The majority of birds in this experiment tended to use the same nest for the subsequent eggs of a clutch whether the previous egg was there or not; and the question arises as to whether the hen is loyal to the nest or to the nest site. By placing a large board under the litter it was possible to move a nest after the hen had laid and to watch her behaviour the next day when she was reintroduced into the pen before laying. In these cases the hens showed loyalty to the site rather than to the nest and if the egg laid the previous day was in the nest in the new site she would roll it to the old site.

It is of interest to compare the behaviour of these domesticated birds with other gallinaceous birds. The nests of other species in this group are also primitive and crudely lined scrapes. Unfortunately observations on actual nest construction are rare. However Watson

(1972) made observations on nest building in the ptarmigan, *Lagopus mutus*. The cock starts making scrapes 2 weeks before the hen lays while the hen generally starts performing this behaviour nearer egg laying. While on the nest the hen gathers nesting material on each side of her and after laying covers her eggs partly or completely with loose grass, lichen, moss or other vegetation (MacDonald, 1970). Several aspects of this behaviour are similar to the behaviour of the domestic fowl, particularly the primitive nature of the nest and the early involvement of the cock. (In extensive conditions the domestic cock of some breeds will lead the hen to potential nest sites and even scrape and sit in them.) Also similar is the gathering of nesting material while sitting. In those species of the family Pheasianidae in which the nest is scantily constructed and relatively open to predators the nest site is of great importance, for hatching success may largely depend on concealment and camouflage. It is of interest therefore to recall that, to the domestic hen, the nest site is, despite domestication, still more important than the nest. However, as yet we do not know what stimuli attract her to a particular site.

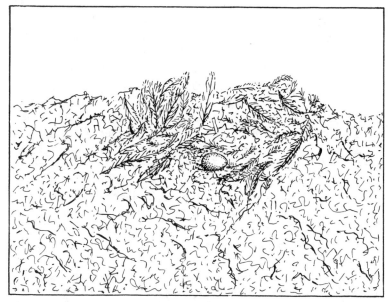

Fig. 6.1 A nest built by a hen in a pen with wood shavings and feathers scattered on the floor. (From Wood-Gush, 1975.)

Incubation and brooding behaviour

These two terms are sometimes used interchangeably which confuses what may be two processes controlled by different physiological mechanisms. Following the example of Eisner (1960) incubation will be used to describe the behaviour associated with sitting on the eggs up to the time of hatching while brooding behaviour will be used in connection with the care of the chicks. Early work (Riddle *et al.*, 1935) indicated that prolactin was responsible for the onset and control of both these components of parental behaviour. Recently Lea *et al.* (1981) examined the levels of prolactin in the plasma of hens as they started to incubate their eggs and also while incubation was in progress. Prolactin levels rose markedly before incubation began and reached a maximum 3–5 days after the start of incubation. However such a correlation does not necessarily prove that there is a causal relationship; other factors may start it. Conflicting evidence was produced by Opel and Proudman (1980) who failed to induce incubation by injections of mammalian prolactin into hens known to have been 'broody' on previous occasions. Thus it is still doubtful if prolactin is responsible for the onset of incubation although it may be responsible for maintaining it and for the onset and maintenance of brooding behaviour.

External stimuli may play a role in brooding behaviour. For example, it is possible to make hens broody by cooping them up with young chicks in an otherwise featureless environment for several days (Ramsay, 1953). However this would not mimic the natural way in which broodiness is started and it would therefore be of interest to investigate the relative rôles of internal and external stimuli more analytically. Once broodiness has began it can be maintained for months by the stimuli supplied by the young (Collias, 1952). As we shall also see in the other species, such stimuli are very important.

THE BARBARY DOVE

General description of parental behaviour

Unlike the domestic fowl, in the case of the Barbary or Ring dove (*Streptopelia risoria*) both parents incubate the eggs and feed the young. Climatic factors such as day-length normally bring the male and female into reproductive condition and the male then proceeds to court the female. Following this a nest site is then chosen. In the laboratory conditions under which these birds' reproductive behaviour has been studied, provision of a nesting bowl and nesting material, in the form of hay, serves to start a whole chain of behavioural events related to reproduction. On the first day of the introduction of a pair to a cage the male courts, strutting, bowing and cooing. After several hours they 'select' the nest site, crouching in the nest-bowl and uttering a distinctive coo. The male then gathers nest material and carries it to the female, which stands in the bowl and builds the nest. A week of nest building follows during which time they copulate, and 7–11 days after the start of courtship she lays the first egg which is followed, two days later, by a second. Thereafter both take turns in incubating the eggs which hatch after 14 days incubation. Both parents feed the young squabs with 'crop-milk', a liquid secreted at this period by the lining of the birds' crop. The young are fed until about 2 weeks old and then the parents' interest in them wanes as another reproductive cycle starts.

The rôle of internal and external factors

Analysis of the external and internal factors involved in this cycle has been carried out by Lehrman and his colleagues (Lehrman, 1964). Pairs of males and females were put into cages, one pair to a cage and provided with a nest bowl containing 2 eggs. These birds acted as the controls and they started to incubate the eggs provided after 5–7 days. In another group each pair were separated from each other in the test cage by an opaque division before introduction of the nest and eggs. They tested the hypothesis that handling delayed incubation (see Fig. 6.2). They too sat 5–7 days after meeting and being exposed to the nest and eggs. In a third group of birds, each pair was kept together for seven days and provided with a nest bowl and hay during that period. They sat as soon as eggs were offered. In a fourth group the birds of a pair were kept together for seven days before a nest and eggs were

Fig. 6.2 (a) Readiness to incubate was tested with four groups of eight pairs of doves. Birds of the first group were placed in a cage containing a nest and eggs. They went through courtship and nest-building behaviour before finally sitting after between five and seven days. (b) Effect of habituation was tested by keeping two birds separated for seven days in the cage before introducing nest and eggs. They still sat only after five to seven days. (c) Mate and nesting material had a dramatic effect on incubation-readiness. Pairs that

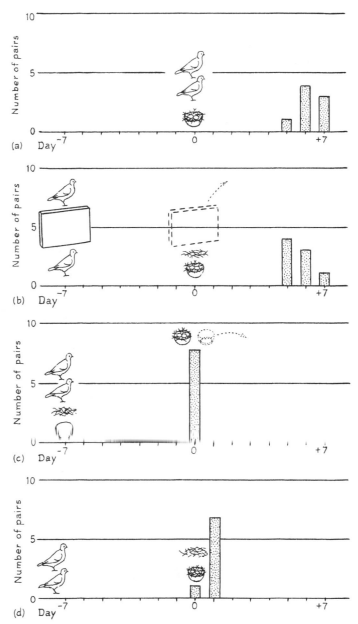

had spent seven days in courtship and nest-building sat as soon as eggs were offered. (d) Presence of mate without nesting activity had less effect. Birds that spent a week in cages with no nest bowls or hay took a day to sit after nests with eggs were introduced. (From Lehrman, 1964.)

introduced. They took a day to start incubation. Therefore it seems that the presence of the male and the nest act synergistically to produce incubation behaviour. Similarly incubation behaviour can readily be elicited in birds of either sex treated with progesterone. Thus there appears to be an effect of the external stimuli on hormone secretion in these birds which as we shall see also seems to happen in other cases. Exogenous prolactin had a very marked effect on crop growth but produced incubation behaviour in less than half the birds treated, and so in this respect the Barbary dove does not differ greatly from the domestic fowl.

In summary then, it seems that the presence of the male, augmented by the presence of the nest bowl and nesting material, induces secretion of the gonad-stimulating hormones (LH and FSH) by the female's pituitary. These hormones in turn stimulate the secretion of oestrogen and progesterone so that incubation behaviour readily occurs and that stimuli from the eggs, once incubation has begun, stimulate prolactin secretion. Progesterone also appears to be the hormone that initiates incubation in the male (Cheng, 1975). Further experiments (Lehrman, *loc. cit.*) indicated that the presence of squabs has an even greater effect on prolactin secretion.

The Barbary dove and domestic fowl show many interesting similarities and contrasts. The most striking difference is of course the rôle of the male in the Barbary dove but the difference may not be absolute for the occasional female housed with a castrated male will build a nest by herself (Dr Carl Erikson, personal communication). Further research on incubation and broodiness in the hen may reveal further similarities.

THE EWE

The physiological basis of maternal behaviour

The pre-parturient ewe moves away from the flock and when the amniotic fluids are exuded, she sniffs and licks the area where they fall, sometimes scraping it. It has been suggested that this behaviour helps her to recognize her own lamb which, when born, is licked assiduously. Parental behaviour in the ewe is characterized by her maternal responsiveness and by the formation of a bond between the ewe and the lamb.

Poindre and Le Neindre (1980) have described the psychophysiological bases of the ewe's maternal behaviour in the following way. The responsiveness of the ewe is dependent mainly on her hormonal status and on the neural stimulation provided by pregnancy and parturition. Over the last two weeks of pregnancy ewes are attracted to new born lambs but such attraction is also seen in some ovariectomized ewes which show maternal behaviour when presented with a new born lamb three times in 24 hours. Full maternal behaviour including licking of the young, the emission of low-pitched bleats, acceptance at suckling and the establishment of a discriminating bond can all be induced by oestradiol-17β and progesterone treatment without parturition.

The maternal reactions of intact ewes have been studied in relation to their hormonal status over the reproductive cycle. About 30% of ewes will show maternal behaviour at oestrus when they have high oestradiol levels and rising progesterone levels. This proportion increases to about 40% some twenty days before parturition but in mid-pregnancy maternal behaviour can hardly be elicited at all, although progesterone levels are increasing at that time. The relationship between the behaviour and the hormones are shown in Fig. 6.3, but Poindre and Le Neindre (*loc. cit.*) conclude that in the intact ewe endogenous oestrogen appears to be more closely related to maternal behaviour than endogenous progesterone (Fig. 6.3).

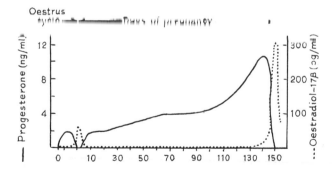

Fig. 6.3 Changes in serum levels of progesterone and oestradiol-17β from oestrus to parturition in the ewe. (From Poindre and Le Neindre, 1980.)

The ewe–lamb bond

After birth, the lamb is usually licked by the ewe; it is thought that this enables the ewe to recognize her lamb and to discriminate against strange ones. Smith *et al.* (1966), on the evidence with very small numbers in experiments, in which the lamb or lambs were removed from the ewe and strange lambs substituted, postulated that 20–30 minutes contact is necessary for the ewe to form a bond with her own lamb. Furthermore there is a sensitive period in which she can develop this. Their work has shown that ewes separated from their lambs for 8 hours from birth, and removed from the natal pen so as to avoid reinforcement from the amniotic fluids, readily accept their lambs. In another separation experiment Alexander and Williams (1966) reported good acceptance of the lamb by Merino ewes following a 24 hour separation immediately after birth, although it is not known if the ewes had formed a discriminating bond with their lambs. However Poindre and Le Neindre (*loc. cit.*), also working with Merinos, reported that a 24 hour separation resulted in six out of eight ewes rejecting their lambs.

Teat-seeking behaviour and suckling behaviour

Very often the lamb begins its teat-seeking behaviour by exploring backwards and forwards along the dam's flank (Bareham, 1975). The ewe may assist by adopting an appropriate orientation and grooming behaviour, but neither of these is essential for the lamb to find the teat (Alexander and Williams, 1964). The actual releasers that guide the new-born lamb to the teat are unknown, but once it has 1–2 hours post-natal experience the lamb appears to use visual cues at close quarters (Bareham, *loc. cit.*). Hunger does not appear to be the factor motivating teat-seeking in the newly born lamb as the introduction of 600 cm^3 of ewe's milk into the stomach does not suppress teat seeking completely, and suggests that there is some other motivating stimulus (Alexander and Williams, 1966).

Once suckling behaviour has developed, it occurs once or more per hour in the first week and decreases to once in six hours in the twelfth week. Normally the lamb passes in front of the ewe which smells it and then stands to be suckled. Alien lambs can get access to the udder by sucking from between the back legs once the ewe is suckling her own lamb. Such behaviour may be seen in intensive conditions (Poindre and Le Neindre, *loc. cit.*).

Ewe–lamb recognition

Several experiments have been carried out to discover how the ewe recognizes her own lamb. Casual observations on the greeting of a ewe and her lamb after separation indicate that bleating often brings them into contact and as they come close together the ewe sniffs the lamb before standing to let it suckle. This suggests that auditory, visual and olfactory cues are used in recognition, but at different distances. In one experiment Alexander and Shillito (1977) tested multiparous ewes with their lambs 3–14 days after birth, in an area 14 × 8 m after a 4 hour separation. Different parts of the lambs' bodies were blackened (see Fig. 6.4). Ten ewes and their lambs were allotted to each treatment. The ewes dodged their lambs when their heads were blackened or the lambs were wholly black while only one out of ten ewes dodged her lamb which had a blackened rump and only two out of ten dodged their lambs which had blackened ears. However some

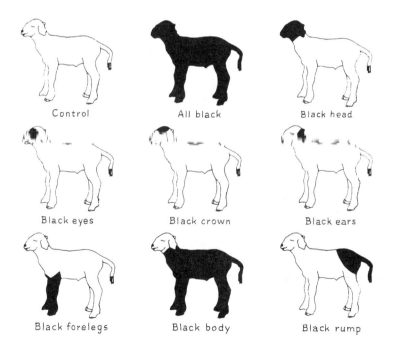

Fig. 6.4 The regions of the lamb's body that were blackened in the experiment by Alexander Shillito (1977.)

ewes were upset by the blackened rump of their lambs when they were at close quarters for suckling.

In another experiment Alexander and Shillito (1977) tested ewes for the relative importance of sight, sound and smell in the recognition of their lambs. Ewes were separated from their lambs for 2–4 hours after testing and then placed singly in a corridor in which three lambs were individually caged, one lamb being that of the ewe being tested. Six treatments were used: the lambs were made (a) silent by means of a local anaesthetic on the vocal chords, (b) motionless by a general anaesthetic, (c) invisible by means of screens round the cage. The other treatments were a combination of immobility, silence and invisibility, and finally a control untreated group. The ewes made most mistakes when they could neither see nor hear their lambs. When the ewes could see, hear and smell their lambs they made very few mistakes while those which could hear and smell their lambs had intermediate scores. The results suggest that smell is not very important at distances of 0.25 m or more; the former distance being the minimum which separated the ewes from their lambs in these tests. In an earlier experiment Morgan *et al.* (1975) used ewes that had been deprived surgically of their senses of smell, taste and touch (oral) and hearing, months before lambing. They concluded that olfactory cues were the main ones employed by the ewe. Alexander and Shillito suggest that the discrepancy between the two findings is due to the fact that the ewes used by Morgan *et al.* had been operated on several months before the birth of their lambs, and thus had learnt to use compensatory senses.

Lambs apparently recognize their dams at a distance by auditory cues and by the age of 7 days know her by her calls (Arnold *et al.*, 1975). They also use visual cues at a distance (Shillito and Alexander, 1975) and may at close quarters use olfactory cues, for the latter authors report that lambs sometimes appeared to reject strange ewes by scent, having approached her udder.

THE FEMALE RAT

A general description of maternal behaviour

Gestation in the rat lasts 22–23 days and as pregnancy proceeds the female spends more time licking her nipple lines and her pelvic and genital areas. About 24 hours before parturition she constructs a nest and after the birth of each pup she eats the placenta and licks the pups, concentrating on the uro-genital area. With the advent of the young she shows full maternal behaviour which is characterized by three main components: Firstly, nest-building in which the pre-parturient nest is made compact. Secondly, nursing in which she crouches over the young, exposing her teats and as the pups nuzzle her fur she changes her position to adjust the pups attachment to the nipples. Thirdly, there is retrieval in which any pup that has strayed from the nest is returned to it. During the first week the pups tend to huddle in the compact nest but in the second week they become more active and stray. However, between the 12th and 16th day post-partum retrieval by the dam declines. In the first week the dam starts most of the suckling bouts but by the third week most are started by the pups and by the end of that week they begin to take food from other sources (Moltz, 1975).

Internal and external controlling mechanisms

This chain of events is controlled by a complex interaction between endogenous factors in the female and exogenous stimuli emanating from the pups. Radioimmunoassay techniques show fluctuating levels of oestradiol, progesterone and prolactin over the course of pregnancy and parturition (see Fig. 6.5). However interpretation of the causation of the onset of maternal behaviour is far from unanimous. Moltz (1973) state that either the decline in progesterone at the end of pregnancy or the rise in a reduced metabolite of progesterone (20 α-hydroxy-progn-4-en-4-one) sensitizes the CNS to oestradiol and prolactin, which act synergistically to make the female more responsive to stimuli from the pups. On the other hand Rosenblatt *et al.* (1979), while agreeing that the decline in progesterone may facilitate maternal behaviour, emphasize that the rise in oestradiol is the main factor, and deny that prolactin is responsible in any way for the onset of the behaviour. They rest this claim on the fact that an agent that blocks the action of prolactin does not stop the onset of maternal behaviour although it interferes with lactation. In addition, injections of prolactin do not induce maternal behaviour in non-pregnant females. It is of interest to compare the curves showing the relative levels of oestradiol and progesterone in the ewe and the rat, in relation to maternal behaviour (Figs 6.3 and 6.5). However, it is possible to bring about maternal behaviour in virgin rats by exposing

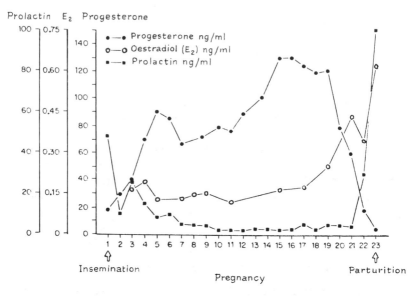

Fig. 6.5 Circulating levels of progesterone, oestradiol and prolactin in the rat from insemination to paturition. ● Progesterone (ng/ml). ○ Oestradiol (E_2)(ng/ml). ■ Prolactin (ng/ml). (From Rosenblatt *et al.*, 1979.)

them continually to young pups for 6–7 days. This treatment also induces the behaviour in hypophysectionized females (Moltz, *loc. cit.*). The maintenance of maternal behaviour appears to be due to the stimuli emanating from the pups. By the third week maternal behaviour declines, probably due to endogenous changes within the females.

SOME PROBLEMS ARISING IN THE MOTHER–OFFSPRING BEHAVIOUR IN FARM ANIMALS

It will be recalled that in monkeys the nature of the mother–infant bond, on the part of the infant, did not depend entirely on the nutritive rewards received from the mother, but on comfort and security obtained from her. Although the young farm animal is not carried by the mother like the infant monkey, the dam will afford it security and shelter in bad weather so that the bond will also depend in part on non-nutritive rewards. Nevertheless, young calves in particular have a very high motivation to suckle and if reared away from the cow will direct the behaviour to other objects, and navel and scrotum sucking can be a considerable problem when calves are grouped together. The motivation for this has not been fully investigated, and we shall consider it again in Chapter 9.

Young heifers and maiden ewes are often frightened of their young and attempt to run away from them, thus delaying the first valuable intake of colostrum. In related wild species this would probably not occur, as there is not the segregation of younger animals from the breeding stock. Thus the young breeding females are likely to have encountered the young of their species before giving birth for the first time. Heifers and ewes in poor condition are often exhausted by the birth and do not stand readily for their new born young, which might then direct its teat-seeking behaviour towards inappropriate objects or, where there are a number of pre-parturient females, might be 'adopted' by another female, only to be abandoned later (Edwards, 1980). Finally, as mentioned in Chapter 3, the calves of multiparous dairy cows often have difficulty in locating the teat due to the pendulous shape of the udder.

POINTS FOR DISCUSSION

1. Discuss how mother–young bonds are formed in sheep and cattle and describe the best husbandry conditions for fostering them.
2. Read the relevant papers cited and list what you consider to be the criteria for good mothering behaviour in the cow, ewe and sow excluding the periods a few days before and after birth; then mention management procedures that might affect these criteria detrimentally.
3. Describe the common methods used in fostering in cattle, sheep and pigs and describe possible ways of improving on them.

FURTHER READING

Donaldson, S.L., Albright, J.L. and Black, W.C. (1972), Primary social relationships and cattle behaviour, *Proc. Indiana Acad. Sci.*, **81**, 345–351.
Edwards, S.A. and Broom D.M. (1982), Behavioural interactions of dairy cows with their newborn calves and the effects of parity, *Anim. Behav.*, **30**, 525–535.
Fraser, D. (1977), Some behavioural aspects of milk ejection failure by sows, *Br. Vet. J.*, **133**, 126–133.

Fraser, D. and Morley-Jones, R. (1975), The 'teat-order' of suckling pigs. I: Relation to birth weight and subsequent growth, *J. Agric Sci. Camb.*, **84**, 387–391.

Gubernik, D.J. and Klopfer, P.H. (eds) (1981), *Parental care in mammals.* Plenum Press, London.

Selman, I.E., McEwan, A.D. and Fisher, E.W. (1970a), Studies on natural suckling in cattle during the first eight hours post-partum. I: Behavioural studies (dams), *Anim. Behav.*, **18**, 276–283.

Selman, I.E., McEwan, A.D. and Fisher, E.W. (1970b), Studies on natural suckling in cattle during the first eight hours post-partum. II: Behavioural studies (calves), *Anim. Behav.*, **18**, 284–289.

Stephens, D.B. and Linzell, J.L. (1974), The development of sucking behaviour in the new born goat, *Anim. Behav.*, **22**, 628–633.

Whittemore, C.T. and Fraser, D. (1974), The nursing and suckling behaviour of pigs. II: Vocalisations of the sow in relation to suckling behaviour and milk ejection, *Br. Vet. J.*, **130**, 346–356.

7
SLEEP AND GROOMING

SLEEP

Like feeding or mating behaviour, sleep is controlled by endogenous factors. It has an appetitive phase in which the animal seeks a suitable site or prepares a habitual site for sleep and assumes a sleeping posture. Finally there is the consummatory phase. However, as will be seen in the ensuing pages, it cannot be easily typified. In general, it is characterized by a number of physiological and behavioural criteria. These include a prolonged period of inactivity, circadian organization (occurring at fairly regular periods over the 24 hour period), raised thresholds of response or arousal, specific sleeping sites or postures and characteristic electro-encephalograms (Meddis, 1975). Indeed sleep may be classified into two types: active and quiet sleep.

Active sleep is characterized by an EEG pattern consisting of low amplitude, high frequency voltage fluctuations almost identical to the EEG pattern of the aroused waking pattern (see Fig. 7.1). In addition, physiological changes are found during this type of sleep (Johnson, 1975). There is absence of tonus in the postural muscles of neck and trunk so that the animal lies flat and there is much rapid movement of the eyes and extremities. These distinctive eye movements have resulted in this type of sleep being called REM (rapid eye movement), or paradoxical, sleep. It is found in all mammals and birds that exhibit any sleep at all. In humans it is correlated with dreaming, and in all species there is difficulty in arousing subjects in this type of sleep.

Quiet sleep on the other hand is characterized by an EEG pattern of high amplitude slow waves (Fig. 7.1). There is a relative lack of movement and animals are more easily aroused from this type of sleep than from active sleep. There are special conditions of autonomic activity (Johnson, 1975). There is high tonus in the urinary and rectal sphincters, dissociation of heart rate and skin conductance. The

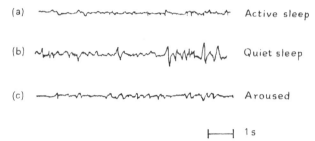

Fig. 7.1 Typical EEG patterns during (a) Active sleep: low amplitude, high frequency voltage fluctuations. (b) Quiet sleep: High amplitude, slower fluctuations. (c) Aroused state: indistinguishable from Active sleep. (From Johnson, 1975.)

postural muscles however are not so relaxed as in active sleep, so that the animal can remain erect to some extent. Quiet sleep which occurs between waking and active sleep is considered to be a phylogenetically primitive trait, for it is found in many classes of animal.

Sleeping habits vary widely, even between closely related species. Not only do the total periods of time per day devoted to sleep vary widely, but the relative proportions of active and quiet sleep also differ. In a survey of data from 39 species from 13 orders, Allison and Cicchetti (1976) found that active sleep was more likely to be found in species with relatively little danger of predation, while larger bodied species spend less time in quiet sleep. However, this does not mean that they spend more time in active sleep as the total time devoted to sleep varies so much (see Table 7.1). It can be seen that many large animals spend little time in sleep; indeed they cannot find safe sleeping sites. Diet can also affect sleeping habits. For example, ruminants must remain in postures which permit rumination to continue (Balch, 1955). Ruckebusch (1972) has analysed the sleep of the cow, pig, sheep and horse, measuring a number of characteristics. Three adult, docile and healthy adults of each species were used. The cows and horses were housed in stalls and the other animals in metabolic cages. Each animal was fitted with chronically implanted electrodes to record the patterns of cortical activity, muscle tone from anterior limb and the dorsal cervical musculature, as well as eyelid movements. Overt behaviour was also recorded together with heart and respiratory rates. In addition to active and quiet sleep, Ruckebusch used another category, drowsiness, which consists of a

Table 7.1 Total sleep time per 24 hours of various mammals. (Adapted from Meddis, 1975.)

Hours	Species
20	Two-toed sloth
19	Opossum, bat
14	Hamster, squirrel
13	Rat, cat, mouse, pig
11	Jaguar
10	Hedgehog, chimpanzee, rabbit
8	Man, mole
7	Guinea-pig, cow
6	Sheep, tapir
5	Horse, bottle nosed dolphin, pilot whale
4	Giraffe, elephant
0	Shrew

mixture of both awake and quiet sleep patterns. The mean total duration of each of the four states for each species is shown in Table 7.2. In each species there was about 10–25% variation between individuals but the intra-individual variation was less than 5%. The progression of the four states is as follows: wakefulness, drowsiness, quiet sleep, active sleep and the changes in the above recordings were registered through this cycle, which was of course repeated many times during a 24 hour period (see Fig. 7.2). As can be seen pigs rapidly lost postural tone, for it decreased during drowsiness and by the time the quiet sleep period started had often completely disappeared. On the other hand, sheep lost muscular tone gradually and some activity persisted in the form of short bursts often in active sleep. Horses showed a gradual loss of muscular tone until the middle of the quiet sleep phase, while cows lost muscle tone abruptly at the onset of active sleep; even then, as in sheep, short bursts of increased muscular tone occurred. In quiet sleep the pigs' eyes were closed, but only partial closure was found at this stage in the other species. However, all species had closed eyes in active sleep. Heart and respiratory rates generally dropped in the transitions from the wakening state to active sleep.

The animals were also tested for arousal to an auditory stimulus. During the drowsy state the threshold of arousal was low in all species and then increased tenfold in quiet sleep. Cattle and sheep ruminate in the wakening and drowsy phases and may even continue to do so in

Table 7.2 The percentage time over 24 hours and over the night devoted to wakefulness and sleep in the horse, cow, sheep and pig. Also shown are the periods of time spent in drowsiness and active sleep, as well as the number of such periods. AW, awake; DR, drowsy; QS, quiet sleep; AS, active sleep. (Compiled from Ruckebusch, 1972.)

Species and time period	Percentage of time				Mean duration and no. of periods	
	Wakefulness		Sleep			
	AW	DR	QS	AS	DR	AS
Horse						
24 h period	80.8	8.0	8.7	3.3	3 min 29 s 33	5 min 13 s 9
Night time (10 h)	52.4	19.0	20.8	7.8	3 min 56 s 29	5 min 13 s 9
Cow						
24 h period	52.3	31.2	13.3	3.1	17 min 57 s 25	4 min 5 s 11
Night time (12 h)	16.0	51.9	25.8	6.3	19 min 40s 19	4 min 30s 10
Sheep						
24 h period	66.5	17.5	13.6	2.4	10 min 4 s 25	4 min 51s 7
Night time (12 h)	49.8	22.9	22.5	4.8	10 min 18 s 16	4 min 51 s 7
Pig						
24 h period	46.3	21.1	25.3	7.3	5 min 50 s 52	3 min 10 s 33
Night time (12 h)	36.5	20.8	32.2	10.5	6 min 30 s 23	3 min 0 s 25

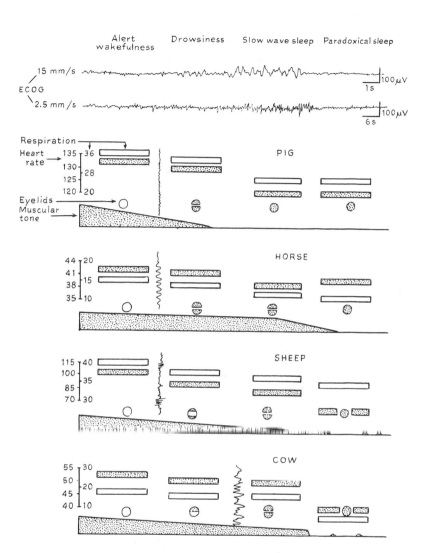

Fig. 7.2 A composite diagram of characteristic electrocorticographic (ECOG) patterns, heart and respiratory rates, muscular tone and degree of closure of the eyelids during the circadian cycle. (From Ruckebusch, 1972.)

quiet sleep, which is of very short duration in their case. Ruckebusch suggests that cattle and sheep have little awareness of the environment when in the drowsy state.

While the relationships between the different physiological measures taken in this study would hold in members of these species in the field, the time devoted to the various stages of sleep would be very different in the field as well might be their proportions relative to one another.

The EEG patterns of sleeping fowls have been analysed by Ookawa and Gotoh (1964), together with simultaneous recordings of muscle tone in the neck and observations of overt behaviour. When most birds tucked the head under a wing, the quiet sleep pattern was found, but this was often suddenly replaced by the active sleep pattern which lasted mostly for about 6 seconds and then reverted to quiet sleep again. Wood-Gush (1959) carried out time-lapse photography on a small group of fowls in a pen under natural light at mid-winter and mid-summer. A roosting perch was available for the birds. In mid-summer the birds which consisted of laying hens and one cockerel went to sleep after a period of preening, while it was still broad daylight. They adopted the sleeping posture of the head tucked under a wing. Over the first 5 hours of sleep in summer and ten hours in mid-winter the birds were generally motionless but thereafter there were frequent changes of posture usually involving moving the head and tucking it under the other wing. No study was made of eye movements. Interestingly, the total proportion of time devoted to sleep did not differ between the seasons and varied from 33–42%. However, in the summer the birds tended to sleep more during the day although the total amount of time devoted to sleep did not differ between the seasons. Recently Amlaner and McFarland (1981) studied the rate of eye movements and periods of eye closure in the herring gull in relation to the bird's posture and threshold of arousal. The experiments were carried out in the field and arousal was measured by administering a series of graded mild electric shocks, either through artificial eggs with metal strips placed in the nest, or a stainless steel grid set on the ground near the nest. The postures of the birds were classified into sleep, rest-sleep and resting postures (see Fig. 7.3). In the rest-sleep posture the rate of eye blinking was greater than in the sleep posture, but it was never seen in birds in the resting posture. Rhythmic eye blinking was found to be correlated with raised thresholds of arousal in both sexes, but males were more easily

Fig. 7.3   The sleep posture (a), rest-sleep posture (b) and rest posture (c) of the herring gull. (From Amlaner and McFarland, 1981.)

aroused than females.

The internal physiological factors that control sleep are as yet poorly understood and sleep can be elicited in animals by stimulating many parts of the brain (Johnson, 1975). The external factors involved will vary from species to species. Many build special nests, others seek out special sites but again we know little about the exact stimuli that elicit the sleeping postures of animals.

Various functions have been assigned to sleep. One is physiological recuperation. Secretion of growth hormone during quiet sleep, promotes amino acid uptake by tissues, as well as protein and RNA synthesis (Oswald, 1980): anabolic processes that apply to both brain and muscles. In humans there are large nocturnal secretions of prolactin, LH and testosterone which are also anabolic. Another hypothesis, which has been put forward by Meddis (1975), suggests that sleep ensures immobility at certain times when immobility would improve the animals' chance of survival, for example, from predation.

GROOMING

Very often grooming is closely connected temporally with sleep or resting. Furthermore, as we shall see in the next chapter, sleep and grooming are commonly found in conflict. A first impression of grooming is that, unlike the various types of behaviour we have considered up to now, it is motivated entirely by external stimuli, but as we shall see, in this chapter and the next, this may not be so. Grooming functions to take care of the body surface and as such is a highly important activity. Furthermore in some species a considerable time is devoted to it. In some of these species it seems to resemble a ritual and, while a bout of grooming might be initiated through cutaneous stimulation, it seems unlikely that it is maintained by that sort of stimulus in such species. Indeed, a detailed study of the grooming behaviour of the herring gull by van Rhijn (1977) has shown that an intricate organization of behaviour underlies it in that species. Learning processes may also be involved. In the chimpanzee, for example, locating a tactile cutaneous stimulus involves learning (Nissen *et al.*, 1951).

Grooming may be affected in different ways by social factors. Allelomimicry is involved in bathing and preening of herring gulls (Van Rhijn, 1977) and the behaviour is generally initiated by the dominant bird. In chickens, dust-bathing also results in allelomimicry; one bird dust-bathing is joined by others. Preening too, may result in other birds doing likewise under certain conditions. In primate societies allo-grooming (the grooming of one animal by another) is very common, and various functions, not mutually exclusive, have been suggested. These have been cited by Seyfarth (1980) as follows: (a) It serves entirely, or in part, to remove ectoparasites. (b) It serves to reduce tension in a population. (c) It helps to establish and maintain close social bonds between individuals. In a field study of Vervet monkeys in Africa, Seyfarth (1980) found that social rank affected the amount of grooming females received. High ranking ones received more than low ranking ones and there was a tendency for females to groom the females of adjacent rank to theirs – a finding that he had previously encountered in other old world monkeys. Allo-grooming is found in cattle (see Fig. 2.16) and horses but the frequency is much less than in primates. However it is of interest to note that, in cattle, allo-grooming is done by animals very close in the social hierarchy (Hafez and Bouissou, 1975). Horses have a limited number of grooming partners (Waring *et al.*, 1975). In pigs allo-grooming is relatively uncommon, but grooming of the ears may be seen between animals that know one another very well. Details of grooming in farm livestock are given in the papers cited below, but the function of allo-grooming in these species is open to speculation as in other species.

POINTS FOR DISCUSSION

1. When going to sleep many animals crowd together. Discuss the advantages and disadvantages of this behaviour.

2. Discuss the causal mechanisms for grooming and describe, if you can, the grooming movements of our common domestic animals.

FURTHER READING

Sleep and resting

Arnold, G.W. and Dudzinski, M.L. (1978), *Ethology of free ranging domestic animals*. Elsevier, Oxford, p. 26–39.

Bell, F.R. (1960), The electroencephalogram of goats during somnolence and rumination, *Anim. Behav.* 8, 39–42.

Kuipers, M. and Whatson, T.S. (1979), Sleep in piglets: an observational study, *Appl. Anim. Ethol.* 5, 145–151.

Reinhardt, V., Reinhardt, A. and Mutiso, F.M. (1978), Resting habits of Zebu cattle in a nocturnal enclosure, *Appl. Anim. Ethol.*, 4, 261–271.

Tyler, S.T. (1972), The behaviour and social organisation of New Forest ponies, *Anim. Behav. Monogr.*, 5, pp. 99–103.

Grooming

Hafez, E.S.E. and Bouissou, M-F. (1975), The behaviour of cattle, in *The behaviour of domestic animals*, 3rd edn (Hafez, E.S.E. (ed.)). Bailliere Tindall, London.

Reinhardt, V. and Reinhardt, A. (1981), Cohesive relationships in a cattle herd *(Bos indicus)*, *Behaviour*, 77, 121–151.

Tyler, S.T. (1972), The behaviour and social organisation of New Forest ponies, *Anim. Behav. Monogr.*, 5, pp. 107–110.

Vestergaard, K. (1979), Normal behaviour of egg-laying fowls. *1st Danish Seminar on Poultry. Welfare in egg-laying cages*. National Committee for Poultry and Eggs, Copenhagen, pp. 11–17.

Williams, N.S. and Strungle, J.C. (1979), Development of grooming behaviour in the domestic chicken, *Poult. Sci.*, 58, 469–472.

8

EXPLORATORY BEHAVIOUR

DESCRIPTION

In previous chapters we have discussed appetitive behaviour in which an animal searches for a particular goal to fulfil a need. The behaviour is flexible, containing a high proportion of learnt responses, and is in a sense exploratory. However, there is another type of behaviour to which the term exploratory is applied. In this the animal appears to be exploring for the sake of exploring and satisfying its curiosity. In practice the two types of behaviour can only be distinguished by detailed study of the sequence of behaviour. Exploration appears to be increased in rats by novelty, complexity and spaciousness (Hinde, 1970), and as novelty often elicits fear, exploration is usually accompanied by some fear. Most studies on exploration have been carried out on rats kept under laboratory conditions and in the case of these animals it has been found that rats in a Skinner box will learn to press a lever to hear a click or to gain access to a running wheel. It seems that rats tend to behave so that, within a given period of 24 hours, they may achieve a certain level of excitation of the CNS via their external sense organs (Barnett, 1975). If one sensory modality is destroyed then exploration is increased to make up for the decreased sensory input. For example, blinded rats explore more (Glickman, 1958). Generally it is suggested that animals like to have a certain level of sensory input; if it is too high they will try to reduce it, if it is too low they will increase their exploratory behaviour (Berlyne, 1960).

A few studies have been made with farm livestock, some of which have studied exploratory behaviour directly while in others there have been implications relevant to exploratory behaviour. Murphy and Wood-Gush (1978) studied the behaviour in strange environments, of two strains of chickens (see also Chapters 9 and 14), one a very flighty strain and the other docile. Their experiment revealed the difficulty of measuring exploratory behaviour and the relationship between novelty, fear and exploration. When the strange environment differed from the bird's home environment in a number of respects, all birds, regardless of age or strain, appeared to be very frightened and there were very few behavioural differences discernible between the strains. However, when the strange environment differed only relatively slightly from the home environment, striking behavioural differences were found. The birds were released singly from a starting box into a strange empty pen. The flighty strain birds emerged from the starting box after a mean time of 449 ± 109 s while the docile strain took on average 2245 ± 598 s to emerge. Furthermore, on emergence the flighty strain birds dashed across the pen before stopping to investigate it, whilst the docile strain birds, which had explored it visually before emergence, came out walking slowly and cautiously. The area of the pen was divided into squares and it was found that the flighty birds entered more squares but that their movements were less relaxed than those of the docile birds. Thus even within a species it may be seen that the measurement of exploratory behaviour presents difficulties.

Mackay and Wood-Gush (1980) examined the behaviour of suckler calves when in a novel environment and when presented with a novel stimulus in that environment. The calves had previously been either loose-housed in groups or individually penned. The latter investigated more areas in the new environment and were more likely to approach the strange object than the loose-housed calves. They also appeared to be more frightened. These results suggest that the barrenness of an animal's environment or its restriction enhances its responsiveness to novelty. Stolba and Wood-Gush (1980, 1981) found this to be the case in young piglets. They observed the responses of piglets in 4 different types of environment to a hanging tyre. In each environment 10–14 growing pigs aged 45–6 months were exposed to the stimulus for 80 minutes or more on several days. The environments were:

1. An indoor Danish pen with a concrete, partially slatted floor.
2. A straw-bedded open-front pen.
3. An enriched open-front pen with a bedding area, a rooting area, and several environment features.
4. A large semi-natural wooded enclosure where the young pigs were in families with adults and sub-adults.

The barer the environment, the more strongly the group reacted towards the stimulus. Strong reactions to the tyre were seen in the first two environments for more than 80 minutes before the tyre was neglected by the piglets for even 30 seconds, while in the semi-natural environment no attention was paid to the tyre in any trial after 10 minutes, and in the enclosed pen in any trial after 30 minutes. The mean proportion of animals in a group reacting to the stimulus during the period of group reaction was significantly greater the barer the environment. Furthermore in the barer environments the response of the pigs to the sudden opening of an umbrella was much greater than in the other environments, suggesting, as in the calf experiment, that in impoverished environments arousal in pigs is much higher when they are presented with a novel stimulus.

In very bare environments such as found in laboratory or intensive agricultural systems animals may perform many different sorts of behaviour, which have no clearcut objectives. Wood-Gush and Beilharz (1982) reported that piglets in a flat deck cage slept more than piglets in an enriched cage. Some animals perform stereotypies in which the same behaviour patterns are repeated over and over again, while others may perform operant responses for apparently trivial rewards. We shall return to discuss these cases later when we consider animal welfare (Chapter 14) but at present their function appears often to be to alter the level of sensory input, and in this respect they resemble exploratory behaviour, although the factor of curiosity is lacking. In animals living in the wild, exploration is unlikely to serve this function for the animal will continually be exploring features of its environment and its main function will be to store information about the details of its environment through latent learning (Lorenz, 1981).

The earlier writers (Berlyne, 1960) envisaged that exploratory behaviour was controlled by the reticular formation, which is a diffuse network of fibre tracts in the brain important in controlling the level of arousal of the animal. Sensory information is carried by specific sensory pathways to the relevant parts of the brain but simultaneously the information is carried to the reticular formation which exerts a modulating influence on those areas, inhibiting any action if the information is irrelevant, and stimulating them if necessary. More recent work has shown that stimulation of specific nuclei can affect the exploratory behaviour of rats (e.g. Weingarten and White, 1978; Rompré and Miliaressis, 1980). However the control of the integration of the behaviour remains obscure.

CONCLUSIONS IN RELATION TO ANIMAL HUSBANDRY

While it is evident that exploratory behaviour is not a motivational system like feeding or drinking behaviour, with measurable deficits and clearcut consummatory acts, or like sexual behaviour with obvious consummatory acts, it nevertheless is a specific motivation system. Indeed, Lorenz (1981) points out that although the fixed action patterns that compose it are found in other motivation systems, in exploration they follow one another with such rapidity that they must surely be under a special motivational system at that time. As stated earlier, in wild animals it is a vital component of their behaviour enabling them to gain very detailed knowledge of their environments. During the course of domestication there has not been active selection against it and so our modern farm livestock will still retain the propensity to explore and to be curious about their surroundings. Deprivation of this opportunity (as shown above in pigs) leads to over-reaction, and it is possible that in dull environments a real need develops. This we shall discuss again in Chapter 14.

POINTS FOR DISCUSSION

1. Compare appetitive behaviour and exploratory behaviour.
2. Discuss some possible advantages that a 'sense of curiosity' might bring to hill sheep or cattle kept extensively.

FURTHER READING

Stevenson, M. (1982), The captive environment: its effect on exploratory and related behavioural responses in wild animals, in *Exploration in Animals and Humans* (Archer J. and Birke L., eds). Van Nostrand Reinhold, London.

Toates, F. (1982), Exploration as a motivational and learning system: a (tentative) cognitive–incentive model, in *Exploration in Animals and Humans* (Archer J. and Birke L. eds). Van Nostrand Reinhold, London.

Wood-Gush D.G.M., Stolba A. and Miller C. (1982), Exploration in farm animals and animal husbandry, in *Exploration in Animals and Humans* (Archer J. and Birke L., eds). Van Nostrand Reinhold, London.

9
CONFLICT AND THWARTING

In ethology the term 'conflict' is used to mean conflict between behavioural tendencies, rather than conflict between animals. Such conflict between motivational systems will arise frequently in animals. For example, an animal may want to approach a source of food, but be frightened of a dominant animal which is already there, so that it is then in a conflict between approach and avoidance. Another type of conflict may arise when initially only one motivational system is aroused, e.g. a hungry animal may see a food source but find it to be inaccessible and thus be thwarted. It is uncertain whether thwarting leads to activation of a second motivational system, such as avoidance, which conflicts with the first, but the behavioural consequences to this type of thwarting are generally similar to those seen in a conflict between two motivational systems.

TYPES OF CONFLICT BEHAVIOUR

Displacement activities

A common response in conflict is a displacement activity which may be defined as an apparently irrelevant behaviour pattern. In many animals common displacement activities are grooming movements. However these movements differ from normal grooming movements by appearing to be exaggerated in form, or more flurried, and often incomplete. Displacement preening in the chicken appears to be more flurried than the preening invariably seen when the birds have woken up or are about to go to sleep on the roost. A film analysis by Duncan and Wood-Gush (1972a) of displacement preening by hens showed that a bout of preening (the period when the feathers were in the bill or

Table 9.1 A comparison of the mean duration of preens by hens when frustrated and not frustrated. The duration is measured by the number of film frames. Also shown are the values from *t* tests and the levels of significance of the differences (*P*). (From Duncan and Wood-Gush, 1972a.)

Bird	Situation	Duration	t	P
White	Control ($n=39$)	42.76 ± 8.05		
	Frustrated ($n=92$)	28.90 ± 3.09	1.98	<0.05
Blue/white	Control ($n=33$)	51.15 ± 7.80		
	Frustrated ($n=19$)	19.26 ± 3.87	2.97	<0.01
Blue/pink	Control ($n=33$)	71.42 ± 8.13		
	Frustrated ($n=69$)	34.33 ± 3.46	4.88	<0.001

a similar but incomplete movement without feathers in the bill) was significantly shorter in displacement preening than in normal preening, giving it an appearance of being flurried.

Originally, when displacement activities were first described, their causation was thought to be due to either an 'after discharge' of energy after intense excitation in the absence of appropriate stimuli, or the sparking-over of energy due to the conflict of two incompatable motivational systems to a third motivational system and thus producing what seemed to be an irrelevant activity (Tinbergen, 1951). These hypotheses were later replaced by the 'disinhibition' hypothesis which suggested that the two conflicting tendencies inhibited one another and in doing so allowed a third motivation system or tendency to be expressed, i.e. to become disinhibited (Van Iersel and Bol, 1958). That this third behaviour pattern is often concerned with grooming was explained by the suggestion that the stimuli which cause grooming are always present but are usually ignored because of other stronger stimuli affecting the animal. If, however, the feathers of a bird in conflict are wet so that they present a stronger stimulus than usual, then the bout of displacement preening is longer (Van Iersel and Bol, *loc. cit.*). Backing for this hypothesis came from observations of van Iersel and Bol which revealed that displacement activities tended to occur when the conflicting tendencies are of the same strength.

Andrew (1956) pointed out that many of the displacement activities in birds resemble their toilet activities and that these could be connected with changes in the autonomic nervous system due to the conflict. For example, autonomic nervous activity could lead to pilo-erection which could elicit grooming. Such a view naturally questions the irrelevancy of the displacement activity.

A variant of the disinhibition hypothesis has been put forward by McFarland (1965) who makes a distinction between the two types of conflict; that due to the conflict between two tendencies and that due to the thwarting of a single behavioural tendency. In a conflict situation, i.e. an approach–avoidance conflict, the animals he worked with, thirsty Barbary doves, tended to show ambivalence, walking backwards and forwards. When they stopped they paid little attention to the environment, their stance generally revealing ambivalence, looking at the goal with a stance indicating readiness to withdraw which McFarland labelled SAV. In thwarting situations on the other hand, the birds tended to pay more attention to the environment and when stationary, stood with an attentive posture (SAP). He suggests that, in a conflict between two tendencies, any displacement activities that occur may be brought about by disinhibition, but in thwarting situations it is a switch of attention during SAP that leads to displacement activities. The animal switches its attention from the inaccessible goal to another object and directs the behaviour towards that object. For example, a thwarted hungry animal may switch its attention to its own plumage from the inaccessible food and start to preen, but because the outcome of the displacement activity is very different from the expected outcome of the feeding behaviour it will soon stop, e.g. feathers in the mouth are very different from food in the mouth. However the behaviour is likely to occur again frequently.

As stated in Chapter 8, stimuli relayed to the brain result in two types of responses occurring. One is related specifically to the stimulus so that the appropriate loci of the brain are activated and the other involves the activation of the reticular formation which is a diffuse series of fibre tracts. From the reticular formation fibres go to the cortex, and through these non-specific pathways the reticular formation can affect the level of arousal in the animal (Manning, 1979). It has been suggested that non-specific arousal involving the reticular formation may be responsible for some displacement activities. Bindra (1959) has suggested that under these conditions a behaviour pattern with a high habit strength, i.e. one that is frequently used would tend to occur as the displacement activity.

Sleep is a common displacement activity in animals and Delius (1967) found that by stimulating the forebrain and brain stem of gulls, preening, sleeping and yawning, which are displacement activities in the repertoire of gulls, often appeared together when particular loci in the brain were being stimulated. Loci that did not produce preening on stimulation did not produce these other behaviour patterns and Delius suggests that these patterns reflect the activation of a more or less unitary system leading to 'de-arousal'. According to this theory the appearance of these behaviour patterns in conflict situations is due to a mechanism for arousal homeostatis brought into action by the arousal resulting from the conflict so that the displacement activities would function to restore the animal to a calmer state.

None of these theories is entirely satisfactory. None have fully considered the exaggerated nature or frequent incompleteness of many displacement activities although this factor could be accommodated in McFarland's hypothesis. Possibly no one mechanism controls all the displacement activities of a species. Not much attention has been paid to differences in displacement activities shown by individuals, but experience has been shown to be important: Red Jungle fowl cocks, that had been individually fed different coloured grains, when displacement pecking during a combat tended to peck at scattered grains of the same colour as their food grains (Freekes, 1971). The external conditions can also influence the type of displacement activity, e.g. fighting turkey toms might show displacement drinking if water is at hand or might show displacement feeding if food particles are near by (Räber, 1948).

The high probability that many displacement activities may be direct consequences of physiological changes in the animal, together with the evidence of action by external factors, has made some ethologists ask if displacement activities are a special type of behaviour pattern as they were originally thought to be. Nevertheless the frantic and exaggerated performance of many displacement activities does seem to separate them from the normal overt responses due to the physiological changes thought to be involved.

Conflict may also lead to other specific types of behaviour patterns:

Redirected activities

Here the motor patterns appropriate to one of the conflicting tendencies are shown but are directed towards an object other than that which initially elicited them, e.g. in herring gulls, pecking during aggressive encounters is often redirected at objects in the environment.

Intention movements

An intention movement may consist of the initial phases of movements or movement sequences of a behaviour pattern. For example, the take-off leap of a bird before flying consists of 2 phases; first the bird crouches, withdraws its head and raises its tail, and then reverses these movements as it springs off. In conflict a bird may repeat these movements for much of the duration of the conflict, or repeat a modified form of them.

Alternation

The animal may alternate between the two conflicting tendencies of say approach and avoidance when wishing to approach a source of food but for some reason is frightened to do so.

Ambivalent behaviour

Sometimes intention movements appropriate to the two tendencies are combined into a single pattern which contains components appropriate to both tendencies. A good example of this is the posture of a cow, or steer, presented with an unusual stimulus. Generally the relative positions of the head and neck express curiosity but the legs and weight of the animal are positioned to enable a sudden withdrawal (Fig. 9.1).

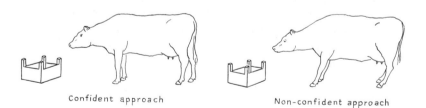

Confident approach   Non-confident approach

Fig. 9.1 The left-hand figure shows a heifer confidently approaching a novel object and the right-hand figure shows a heifer approaching a novel object in a non-confident mood. Note the difference in the position of the legs in the two moods. (By kind permission of Candace Miller.)

Compromise behaviour

This is similar to ambivalent behaviour but instead of a compound behaviour pattern consisting of parts of the two conflicting tendencies, only one pattern is shown which can express both tendencies. The waltz of the cockerel performed in agonistic interactions seems to be a compromise of approach and withdrawal. The path traced, for example, is circular.

Fixations and stereotyped behaviour

When animals are presented with an insoluble problem such as a visible but inaccessible goal that is desired, they will in time begin to perform stereotypies, repeating the same behaviour patterns over and over again. Duncan and Wood-Gush (1972) studied the onset of a stereotypy in hens. The birds were divided into groups some of which were trained to varying degrees to expect food when put in the test cage, while others had no expectation of being fed there. Eventually the birds were put singly into the test cage with the usual food in the usual position but it was now covered by a pane of clear glass. In addition to the variable of expectancy was added the variable of hunger, so that the groups now varied from one group that was not hungry and had no expectancy of finding food in the test cage, through groups that had varying degrees of expectancy and hunger to a group that was both very hungry and in a state of high expectation (VHHE group). The birds with little expectation and little or no hunger preened during the test, while the VHHE group birds performed stereotyped escape movements. The birds in the intermediate groups were very interesting in that they began to give alarm calls and tried to escape, suggesting that they found the situation aversive, and suggesting that the stereotypies in the VHHE birds acted to reduce the anxiety engendered by extreme frustration or thwarting. The stereotyped escape behaviour in the VHHE hens continued in the test cage even when the pane of glass was removed; the birds might peck at the food but carry on with the stereotyped escape behaviour and not feed. In other words, the behaviour became fixated. When given a tranquilizer that would not affect their appetite they still continued with the stereotypy (Duncan and Wood-Gush, 1974). However the frequency of the stereotypy was reduced. In a second experiment administration of the drug before frustration delayed the onset of stereotypies. As the drug used is well known for its fear-reducing

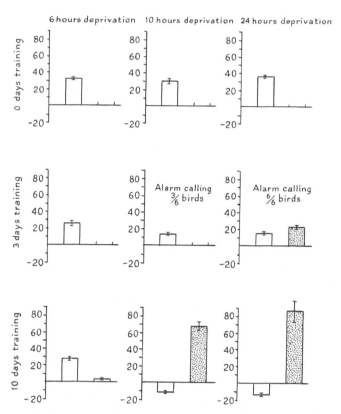

Fig. 9.2 The behaviour of hens under varying degrees of frustration when prevented from reaching food that was easily seen. Some were hungry (see top from left to right) and some had a high expectancy of being fed (see left hand column from top to bottom). The white columns show the increase or decrease in the number of preens and stereotyped escape movements (shaded columns) in the frustrating situation compared with the measurements of these behaviour patterns in the control situation in which they were not hungry and there was no food present. An increase is shown by a column arising above the x-axis and a decrease by one falling below it. (From Duncan and Wood-Gush, 1972b.)

properties these results suggested that the stereotypy in this case was initially caused by fear or aversion but that it was maintained by some other factors. Another case of a fixated stereotypy was reported in rats given an insoluble problem (Maier, 1949).

There are also cases of stereotypes which are not fixated. Wood-Gush (1972 and 1975b) studied two strains of laying birds during the hour or so before oviposition in battery cages. The hens of the one strain, a White Leghorn strain, indulged in stereotyped escape behaviour in the hour or so before laying, only pausing to sit for a second or two, several times before laying. Outside this pre-laying period they sat in the cage as much or more than another strain which did not show this stereotypy. However the stereotypy was not fixated, for if provided with a solid floor with litter on it in the same cage they sat during the pre-laying period.

Several cases of non-fixated stereotypies have been described in animals kept in barren and unvarying environments and this behaviour has been circumvented by enrichment of the environment. Canaries in cages sometimes spot-peck or indulge in stereotyped flying, and Keiper (1970) showed that these stereotypies could be reduced by enriching the environment (Fig. 9.3). Calves housed in individual stalls also show stereotypies (Kiley-Worthington, 1977). Similarly tethered sows perform stereotypies such as bar-biting which Fraser (1975) found could be reduced by giving the animals straw. The stereotyped pacing or running seen in some caged zoo animals has been suggested by Morris (1964) to have arisen in some cases, because

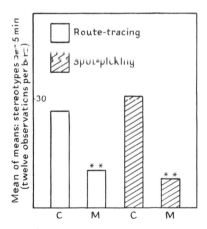

Fig. 9.3 The frequencies of stereotyped behaviour patterns for canaries in a standard cage with and without a plastic mirror. Significant differences between the experimental (M) and control (C) conditions are indicated by two stars over a column (1 < 0.01). (From Keiper, 1970.)

the environment in which the animal finds itself has some of the stimuli that release the behaviour normally. However because the environment has only relatively few of these stimuli, the behaviour is shorn of its complexity and is usually reduced to a series of movements, endlessly repeated. A territorial animal in a cage, for example, might find that the cage has some of the properties of a territory that elicit patrolling, but because the cage is small and featureless the complexity of patrolling is lost, and the animal ends up going round and round.

The relationship between fixated and non-fixated stereotypies is still a matter of conjecture, and indeed it is not known whether they have common causal mechanisms. It is quite possible that a barren environment causes thwarting or frustration, as it is mostly called, which thus causes the non-fixated stereotypy. On the other hand the

Table 9.2 The mean score for 14 behavioural categories by tethered sows during five daily 15 minute periods. (From Fraser, 1975.)

Behaviour	With straw		Without straw		Significance of difference
	Mean	SE	Mean	SE	
Posture					
Lie	42.7	3.2	36.0	5.0	<0.05
Stand	33.1	3.2	40.3	4.8	<0.05
Sit	1.6	0.8	1.5	1.1	NS
Stand or sit motionless	0.1	0.1	3.8	1.8	<0.01
Stand or sit with head pushed through bars	0.6	0.1	1.7	0.5	<0.01
Comfort movements					
Rub body on walls	3.6	0.7	5.0	0.9	NS
Scratch with hind leg	2.3	0.7	2.9	0.7	NS
Shake head	1.0	0.2	1.3	0.3	NS
Other activities					
Chew and manipulate straw	24.5	2.8	0		<0.01
Nose and lick bars, floor, trough and chain	3.5	0.9	15.9	2.3	<0.01
Bite bars, trough and chain	3.2	0.9	8.0	2.2	<0.05
Nose and lick neighbour	1.0	0.3	3.1	0.9	<0.01
Bite neighbour	0.2	0.2	0.1	0.1	NS
Bite, nose and lick neighbour's tether	0.8	0.4	1.9	0.7	<0.05

main causation may be a lack of sufficient sensory input. (Baker, 1981), found that the incidence and intensity of stereotypies increased in tethered sows with increasing experience of being tethered. Sows which had spent several pregnancies being tethered not only showed more stereotypies but were less reactive to a novel stimulus, which would suggest that the stereotypies were fixated in that environment. It is likely that they are maintained by the animal's need for a certain level of arousal.

Aggression appears to be closely connected with frustration, but we shall discuss that in the next chapter. However, from what has been said, it can be seen that those species that have been closely studied respond to thwarting and frustration with some rather striking behaviour patterns which can be of potential use in animal husbandry. This point will be considered when we discuss animal welfare.

Vacuum activities. Thwarting may sometimes lead to the animal performing the motivated behaviour in the absence of relevant stimuli. An example of this in chickens was reported by Wood-Gush in the study mentioned earlier in this chapter. The hens of the one strain during the pre-laying period in a battery cage indulged in vacuum nest-building, going through the movements of manipulating non-existent nest material.

Adjunctive behaviour. During the course of some operant conditioning schedules, animals may perform irrelevant activities. This has been called adjunctive behaviour and it appears to share many of the attributes of displacement activities and related behaviour patterns (Falk, 1977; McFarland, 1970).

Displacement activities and other conflict activities, particularly intention movements, are important in that displays are considered to have evolved from them (Tinbergen, 1952). Furthermore conflict behaviour is of importance in assessing the welfare of animals but these points will be dealt with fully in later chapters.

CONFLICT BEHAVIOUR AND ANIMAL CARE

Although a relatively large amount of work has been done on the fowl with respect to this class of behaviour patterns, they have hardly been studied in other domesticated animals. This is indeed an unfortunate omission, for these behaviour patterns are a very useful guide to detecting frustration before its ill-effects reach distressing and uneconomic proportions. Possibly good stockmen have instinctively recognized such behaviour as being correlated with circumstances that upset their stock, but such knowledge is generally not systematic and often good points are balanced by unreliable ones. The use of these behaviour patterns for assessing welfare of farm livestock will be discussed in a later chapter, but household pets are also not free of frustration and a knowledge of their conflict behaviour is required.

POINTS FOR DISCUSSION

1. Can you describe any displacement activities in farm livestock other than the domestic fowl? If so, try to establish the reason for their occurrence.
2. Try to list farm management practices which result in animals showing stereotypies and discuss the possible causes of the stereotypies in these practices.

FURTHER READING

Duncan, I.J.H. and Wood-Gush, D.G.M. (1974), The effect of a rauwolfia tranquillizer on stereotyped movements in frustrated domestic fowl, *Appl. Anim. Ethol.*, 1, 67–76. (This paper has wider implications than those pertaining to the domestic fowl.)

Fox, M.W. (1968), Socialization, environmental factors and abnormal behavioural development in animals, in *Abnormal behaviour in animals* (Fox, M.W., ed.). W.B. Saunders, London, pp. 332–355.

Kiley-Worthington, M. (1977), *Behavioural Problems of Farm Animals*. Oriel Press, London, pp. 74–79.

Lorenz, K.Z. (1981), *The Foundations of Ethology. Springer-Verlag, New York, pp. 242–253.*

Manning, A. (1979), An Introduction to Animal Behaviour, 3rd edn, Edward Arnold, London, pp. 180–187.

10
AGGRESSION AND FEAR

AGGRESSION

Aggression has been mentioned in previous chapters without being defined. Strictly speaking it does not include fighting skill, although if combined with fighting skill it will enhance the social status of the individual animal. It is the tendency to want to inflict damage on another. This definition does not distinguish between intra-specific aggression and inter-specific aggression but here we will only consider intra-specific aggression as it appears that the two types are under the control of different factors, although they may share some behaviour patterns (Archer, 1976, Manning, 1979).

Interest in the underlying causes of aggression in animals is heightened because people studying violence in our urban societies feel that an understanding of animal aggression may help to elucidate the problem. Lorenz (1981) and others have argued that aggression in man and animals is driven endogenously, in such a way that, if not appropriately released, it will inevitably occur. It will 'dam up'. On the other hand Hinde (1969) and others have argued that although endogenous factors are involved, and that these may fluctuate in strength from time to time, external factors and experience are equally, or even more, important. Lorenz's hypothesis is demonstrated by a model of motivation that he has designed which is not limited to the motivation of aggression alone and is intended to explain the motivation of many other systems. So far we have talked of releasers, sign stimuli and key stimuli which release with internal factors a type of behaviour. This is the common view in ethology and stems from the writings of Tinbergen (e.g. Tinbergen, 1951) and is also the view held widely in psychology. Lorenz's theory, while largely overlapping with that of Tinbergen, suggests that in the absence of relevant stimuli the endogenous factors, the action specific energy, may build up and

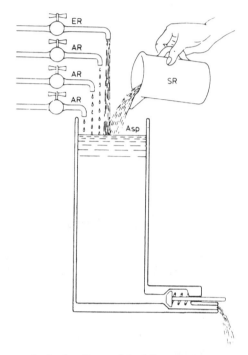

Fig. 10.1 The psycho-hydraulic model of Lorenz (1981). The line Asp symbolizes the present level of action-specific potential. ER represents the source of endogenous and automatic generation of stimuli. SR is the specific releasing key stimulus. The taps AR represent the additional effect of non specific priming or enhancing stimuli. (From Lorenz, 1981.)

eventually be released in the absence of any relevant stimuli (see Fig. 10.1). This type of motivational model is known as the psycho-hydraulic model. It explains vacuum activities (see Chapter 9). These behaviour patterns seem fairly common in birds kept in bare environ-ments but very rare in mammals, and their interpretation is a matter of controversy among ethologists. In the psycho-hydraulic model it is envisaged that energy called action specific energy (ASP) builds up through endogenous factors (ER); in addition there are unspecific stimuli that act to prime the behaviour (AR) and thirdly the relevant key stimulus (SR) acts to release the behaviour. In the model this is represented by the spiral spring which under pressure contracts and allows the action specific energy to activate the behaviour pattern. While this model fits some behaviour patterns it is difficult to envisage the operation of any feed-back mechanisms as discussed in Chapter 3.

The value of any model depends on how much experimentation it engenders so that greater insight may eventually be obtained. In modern production systems farm animals are likely to find themselves in conditions in which the relevant stimuli are absent, whereas this will not be the case for animals in the wild. Hence in the case of the former the validity of Lorenz's model becomes very important when we consider animal welfare.

Internal factors

Testosterone has for long been linked with aggression and injections of testosterone have been shown to raise the social status of chickens (Allee *et al.*, 1939) and aggressiveness in rats, hamsters and red deer (Archer, *loc. cit.*). Other hormones have also been found to control aggression in other species. In starlings, LH is involved (Davis, 1957) and in the African weaver birds, *Quelea*, this gonadotrophic hormone has also been found to be involved in aggression in connection with inter-individual distance (Crook and Butterfield, 1968). Lactating females are very aggressive in many species and it has been suggested that prolactin might be the causal mechanism in these cases. Progesterone may also cause aggression in the female hamster. ACTH may also be involved, in that high levels are correlated with reduced aggressiveness and low levels with increased aggression (see Archer, *loc. cit.* for a detailed review of the action of these hormones and aggression). Different strains within a species may differ a great deal in their readiness to show aggression but the physiology of this phenomenon is poorly understood.

Experience has a very big influence on the level of aggressiveness in an animal. Guhl (1964) reported that although testosterone propionate injections could increase the level of dominance in fowls there was a certain amount of social inertia. By this he meant that previously learnt responses in relation to the birds pen mates had to be overcome before the purely behavioural effects of the hormone became evident. Smith and Hale (1959) likewise found it very difficult to improve the status of hens in 3 very small flocks of 4 birds each, by conditioning. The experimenters attached small saddles to birds which allowed them to administer a small shock to a bird. They proceeded to administer a shock to the dominant birds every time they behaved aggressively to the inferior birds. Soon the dominant birds became non-aggressive but the socially inferior birds took a long time to take advantage of this and increase their status.

Early experience is very important in shaping the level of aggression shown by animals. Those that have been reared in isolation from contact with animals or humans are very aggressive. In mice the absence of litter mates in the pre-weaning stage results in higher levels of aggression. Likewise the absence of peers in the post-weaning stage leads to greater aggression and if peers are missing in both of these phases there is an additive effect (Denenberg, 1973).

Appetitive behaviour

A crucial question for the sociologist seeking further insight into human violence is whether aggression in animals is like primary drives such as hunger or thirst. Is there evidence of appetitive behaviour for aggression? Do animals actively seek situations in which to give vent to their aggression? The evidence for this is not clear-cut, but animals may find aggression rewarding under some conditions. Animals having had a succession of wins fight more readily than others with less experience in winning (e.g. Collias, 1943).

External factors

If in the correct context (e.g. the territory of a male) certain key stimuli can readily release aggression in a number of species. In the stickleback, red on the belly of a model, even a very crude one (see Fig. 10.2) elicits aggression. A cock robin will show more aggression towards a tuft of red feathers in his territory than to a complete bird lacking red feathers. In the case of the domestic hen most aggressive pecks are directed towards the comb of the other hen.

FEAR

The responses given by a mildly frightened animal are identical to those given in withdrawal from a stimulus for other reasons. Therefore fear cannot be as readily defined in terms of the overt behaviour as is done in the case of aggression in which attack and threat are fairly clear-cut. However, as we saw earlier, many threat displays contain elements of fear so an exact appraisal of the degree of aggression may be difficult. This close relationship between

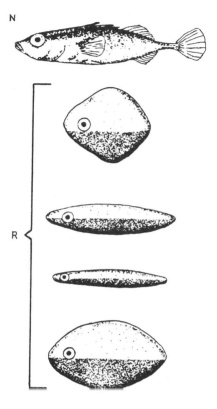

Fig. 10.2 The series of models R each have a red belly and all released attack in male sticklebacks while the model N which resembles a stickleback but lacks a red belly was attacked significantly less. (From Tinbergen, 1951.)

aggression and fear has led to the term agonistic behaviour, to cover both types of behaviour which are inevitably intermingled in many encounters between animals. In the case of fear, different intra-specific strains may give different overt responses although physiologically both strains show signs of fear. Duncan and Filshie (1980) investigated the fear responses of the two strains of fowls mentioned in Chapter 8. The White Leghorn strain on approach by man is extremely flighty whilst the other strain shows little overt response. However the heart rate responses of the two strains as shown by telemetry indicated that the White Leghorn strain returned to its normal rate significantly sooner than that of the other strain. Similarly in the pig, vocalizations which sound fearful are not correlated with the expected physiological parameters (Baldwin and Stephens, 1973). Generally fear is considered as an emotional state and its presence is assessed by the overt behaviour of the animal. But, as discussed above, these responses are difficult to identify unambiguously and therefore several measures of behaviour, as well as the context of the behaviour, are really necessary to assess the presence of fear in animals.

From the classical work of the physiologist, Cannon, it has been known that the presentation of frightening stimuli results in the Emergency Reaction with increased adrenalin or noradrenalin secretion from the adrenal medulla and the subsequent preparation of the animal for fight or flight (see also Chapter 11). Furthermore, high levels of adrenocortico-trophic hormone are correlated with elevated levels of fear (Archer, *loc. cit.*).

The external stimuli that elicit fear are often of high intensity, suddenly presented and have a marked degree of novelty. In addition there are often species-specific stimuli that release fear. Snakes or certain properties of the snake elicit fear in a wide variety of species. In other cases certain properties of the species' main predators may act as releasers of fear. In the case of chickens, anything suddenly thrown into the air will release the aerial predator call. Young chicks of several species will show alarm to models of flying birds if the model has a short neck. Other traits such as the slope and size of wings and tail are irrelevant (Fig. 10.3 and Tinbergen, 1951). Although there has been controversy over this experiment the later evidence appears to support it (Manning, 1979). In addition, patterns resembling eyes are considered to be frightening to a number of species because they resemble the eyes of their predators, which are mainly birds of prey. Also in species with frontally facing eyes, a pair of staring eyes will resemble those of an adversary in a fight and therefore be frightening or very provocative. Even in species with eyes that are not frontally placed a frontal view of the head of a conspecific may also be either provocative or frightening.

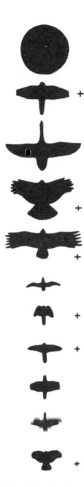

Fig. 10.3 Bird models used for the reactions of various birds from various species of birds of prey. Those marked + released escape responses. Their common characteristic is a short neck. (From Tinbergen, 1951.)

THE RELATIONSHIP BETWEEN AGGRESSION AND FEAR

Aggression is extremely common when animals or man are frustrated or thwarted, so common in fact that a group of American psychologists postulated that frustration is the sole cause of aggression (Dollard *et al.*, 1939). In agricultural animals, aggression has been produced experimentally by frustration in goats and hens. Occasionally fear is produced under frustration and there are a number of other situations which commonly evoke either aggression or fear (Archer, *loc. cit*). These are the ones involving pain, intrusion of an animal's individual distance, the presence of a novel object in a familiar situation and the entering of a strange environment by the animal. Competition which Archer includes under thwarting would of course be the most common of these under farm conditions. Reviewing a large number of cases involving these situations, he concludes that the eliciting of either aggression or fear is due to a discrepancy between what the animal encounters and what it expects. This discrepancy can involve spatial or temporal expectations, e.g. reward at a particular place or time. The outcome will be decided by the stimuli confronting the animal so that a releaser for aggression, for example, will very quickly be effective. He also suggests that as discrepancy increases so it is more likely that fear will increase more rapidly than aggression.

CONCLUSIONS IN RELATION TO FARM LIVESTOCK

The relationship between aggression and fear and the predictable contexts in which they are likely to appear make it difficult to believe that aggression is motivated by a psycho-hydraulic type of system. Early experience, in some species at least, is very important in shaping the level of aggression in the individual, and in others it can be released by very specific stimuli in certain contexts (e.g. the cock robin in the breeding season). A fuller knowledge of the conditions and of the specific stimuli that release aggression and or fear would be of great value for animal husbandry. However, as we shall see when discussing domestication, artificial selection by man has sometimes interfered with such stimuli. It would seem prudent in all types of husbandry to limit isolation and frustration as much as possible; to ensure that the animals are not continually intruding into one another's individual space and to avoid placing animals so that they face one another frontally at close quarters. Generally this can be done by better building design. For example, pigs when feeding outside from discrete piles or scattered food show aggression to one another if the distance between them is less than 2 metres. However, if individual stalls with non-transparent sides that partially hide the neighbour are used, then the animals will feed with very little aggression (Stolba, *personal communication*).

POINTS FOR DISCUSSION

1. Decreasing space for animals generally leads to increased levels of aggression. Explain this increase by means of one or more of the theories put forward in the chapter.
2. Discuss the difficulties in assessing fear.
3. What do you think are the functions of aggression and fear?

FURTHER READING

Aggression

Duncan, I.J.H. and Wood-Gush, D.G.M. (1971), Frustration and aggression in the domestic fowl, *Anim. Behav.*, **19**, 500–504.

Ewbank, R. and Bryant, M.J. (1972), Aggressive behaviour amongst groups of domestic pigs kept at various stocking rates, *Anim. Behav.*, **20**, 21–28.

Hughes, B.O. and Wood-Gush, D.G.M. (1977), Agonistic behaviour in domestic hens: the influence of housing method and group size, *Anim. Behav.*, **25**, 1056–1062.

Scott, J.P. (1948), Dominance and the frustration-aggression hypothesis, *Physiol. Zool.*, **21**, 31–39.

Signoret, J.P. and Bouissou, M.F. (1971), Adaptation de l'animal aux grandes unités. Etudes de comportement. *Bulletin Technique d'Information*, Ministére de l'Agriculture, France, **258**, 367–372.

Fear

Jones, R.B. (1980), Reactions of male domestic chicks to two-dimensional eye-like shapes, *Anim. Behav.*, **28**, 212–218.

Murphy, L.B. (1978), The practical problems of recognising and measuring fear and exploration in the domestic fowl, *Anim. Behav.*, **28**, 422–431.

Murphy, L.B. and Wood-Gush, D.G.M. (1978), The interpretation of the behaviour of domestic fowl in strange environments, *Biol. Behav.*, **3**, 39–61.

11
STRESS

THE PHYSIOLOGY OF STRESS

In animal production parlance, stress is a widely used term generally covering a variety of inexplicable mishaps in stock production. Used loosely in that way the concept loses all value. As defined by Selye (1960), who pioneered the work on stress, it is the consequence of wear and tear in a biological system. He stated (*loc. cit.*) that the physiological response to stressors (noxious agents) such as heat, cold, trauma, infections, nervous stimuli, etc., is generally a stereotyped response in the form of secretion of adrenocorticotrophic hormone (ACTH) which acts upon the adrenal corex to secrete glucocorticoids (i.e. cortisone, corticosterone and cortisol) which convert proteins and fats into glucose and possibly mineralocorticoids (aldosterone and deoxycorticosterone) which affect water retention and the concentrations of electrolytes such as sodium and potassium in the body. Selye originally envisaged that this stereotyped response was given non-specifically to stressors of all sorts. Such a response is immediately beneficial to the animal and he called it the General Adaptation Syndrome (GAS). In the event of continued action by the stressor the initial Alarm State is followed by a Stage of Resistance, but during this phase there is an over-production of the glucocorticoids resulting in atrophy of the thymus gland, a decrease in lymphocytes, plasma cells and eosinophils in the blood so that the immune response is harmed and the animal becomes susceptible to infection. This, the third stage of the animal's response to a prolonged stressor, the Stage of Exhaustion sees the demise of the animal (Fig. 11.1). More recent work has shown that other endocrine organs apparently involved in responses to stressors include the thyroid, the pancreas and the testes or the ovaries. Furthermore in addition to responses by the adrenal cortex under action of ACTH from the

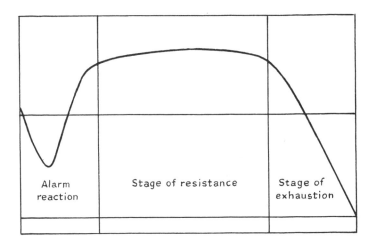

Fig. 11.1 A diagrammatic representation of the general adaptation syndrome showing the triphasic reaction to a stressor. First the alarm reaction then the stage of resistance and, finally, the stage of exhaustion. (From Selyne, 1960.)

pituitary, is the activation of the sympathetic adrenal medullary system so that adrenalin and noradrenalin are produced and the animal shows a syndrome of physiological responses that prepare it for 'fight or flight'. Cardiac output is increased, erythrocytes are discharged more rapidly from the spleen into the blood system; blood is shunted away from the viscera for use in the skeletal muscles; glycogen reserves are broken down to make more glucose available. However very intense sympathetic activity may lead to death through over-stimulation of the heart by adrenalin or through an excessive parasympathetic reaction to the intense sympathetic response, resulting in excessive vagal stimulation which leads to stoppage of the heart. Equally a total lack of stress is likely to be harmful to the animal because it may not be able to give the appropriate physiological response to sustained adrenocortical secretion.

Selye's definition is not entirely satisfactory for it could include the normal process of ageing as a stressor and one proposed by Archer (1979) is more suitable. In this definition stress is the prolonged inability to remove a source of potential danger, leading to activation of systems for coping with danger beyond their range of maximal efficiency. Furthermore as we shall see later, more recent work has shown that the original idea of the response being non-specific to a variety of stressors has to be modified.

Many natural events such as oviposition in the domestic fowl lead to elevated plasma levels of corticosterone (Beuving and Vonder, 1978). The animal may thus be thought of being under stress. However the stressor is short-lived. Similar responses to short term stressors have been reported in other contexts with farm livestock. A good example of a psychological stressor in the form of frustration was shown by Dantzer *et al.* (1980). They taught pigs individually to press a lever to obtain a small amount of food on a continuous reinforcement schedule after 23 hours of food deprivation. When the pigs were proficient, the food reward was withheld and the response was eventually extinguished. Blood samples were taken before the extinction trial

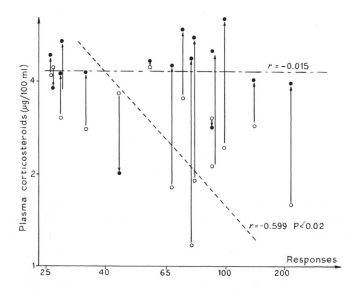

Fig. 11.2 Relationship between the number of panel presses made during an extinction session in pigs trained to obtain a food reward through pressing the panel and their plasma corticosteroid levels. Open circles show for each pig its initial plasma corticosteroid level measured just before the extinction session. The black circles their plasma corticosteroid levels 30 minutes after the end of the extinction session. Response rates were correlated to the initial plasma corticosteroid levels ($r = -0.599$) but not to the final plasma corticosteroid levels ($r = -0.015$). (From Dantzer *et al.*, 1980.)

and again 30 minutes after it. Extinction trails are generally frustrating and the pigs during this trial displayed signs of frustration performing displacement activities if alone, or showing aggression if in pairs. Indeed, in nearly all cases they revealed elevated levels of corticosteroids (see Fig. 11.2). Furthermore these workers had previously shown that the conditioning trials without frustration had not affected the pigs' corticosteroid levels. However it is not only contrived frustration that may lead to elevated glucocorticoid levels but also many husbandry practices. Kilgour and De Langen (1970) studied the cortisol levels of sheep after each of a number of management practices, and their data are shown below.

As mentioned earlier a number of other hormones apart from those produced in the adrenal glands appear to be involved in response to stressors. This widespread involvement was very clearly shown in work by Mason and his colleagues. In several experiments involving either conditioned avoidance or unconditioned emotional responses

Table 11.1 Plasma cortisol levels (μg/100 ml) in sheep after the following management practices. (Basal level, 3.6 μg/100 ml.)

Shearing 5 min	Shearing 10–15	Dipping 5 min	Trucking 90 min	Dog-chasing 5 min
8.3	15.8	5.7	4.6	5.7
7.0	11.6	7.6	6.7	4.3
5.6	7.7	9.1	4.1	6.6
5.8	13.7	7.0	8.3	6.3
6.6	12.4	6.4	8.1	9.6
7.7	10.3	5.3	7.0	16.0*
6.2	10.0	4.2	7.1	5.7
8.9	6.5		5.9	9.1
7.8	15.3		4.8	10.6*
7.8	10.0		7.0	8.7
			8.7	
			7.1	
Av. 7.2	11.3	6.5	6.7	8.5

* Bitten by dog

they showed the involvement of a number of endocrine responses in the male Rhesus monkeys used as subjects. 17-Hydroxycorticosterone (17-OHCS), adrenalin, noradrenalin, plasma TSH and plasma thyroxine elevation occurred simultaneously with decreased levels of testosterone and oestrone. Insulin and growth hormone gave no clear-cut picture. The same endocrine pattern appears to hold for human males when stressed. Furthermore in monkeys the pattern of hormone release is different when they are anticipating something unpleasant and unknown, from when they expect something unpleasant but known. In both cases 17-OHCS, noradrenalin, plasma TSH and plasma thyroxine are elevated but not adrenalin. However in cases in which unknown and unpleasant factors are expected adrenalin levels rise similarly to those of noradrenalin. The same seems to occur in man and one wonders about any similarity in the mental states of the two species under these conditions (Mason, 1975).

Other studies mentioned by Mason have shown that the levels of gonadal hormones are decreased during periods of stress in man as in monkeys but when anger, either overt or repressed, is involved testosterone levels may be elevated. The relative lack of non-specificity of response to stressors is indicated in Fig. 11.3.

In general the picture gained is in keeping with Cannon's notion that during emotional reactions physiological adjustments occur which prepare the animal for the muscular exertion of flight or fight.

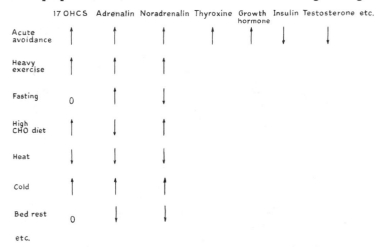

Fig. 11.3 The patterns of hormone responses to different types of stressors. (From Mason, 1975.)

SOCIAL STRESS

For many years it has been known that the numbers of many wild species fluctuate in a cyclical manner. The causation of these fluctuations was unknown; predation, food scarcities, diseases were all considered and found to be inadequate to explain the dramatic collapse in numbers. Following on an earlier lead in the 1930s in which several workers had suggested that psychological factors or 'shock disease' might be implicated in the death of animals, Christian (1950) in light of Selye's work in the 1940s suggested that stress might be the cause. Many of the workers had described animals, such as voles and snow shoe hares, as having convulsive seizures shortly before death and other animals in the same population as being lethargic or comatose. He (Christian, 1955) carried out experiments with laboratory-reared wild mice and Albino mice housing them in cages of a standard size in populations of different sizes. The animals were kept singly from weaning for at least 3 weeks and then put into cages of a standard size, all kept in the same room. Population sizes for the wild mice were 3, 4, 6, 8, 9 and 17 and 4, 8, 16 and 32 for albino mice. All the controls were caged one to a cage. Ample food and water was supplied. The body weights of the wild mice in the different populations did not differ from those of the controls but their adrenal glands were significantly heavier than those of the controls. Thymus gland weight, preputial gland weight (indicative of androgen activity in mice) and testes length were smaller than those of the controls in mice from all populations of all sizes (see Fig. 11.1).

The albino mice generally had lighter body weights than their controls and heavier adrenal glands. On introduction to the experimental cages there had been severe fighting for 1–3 hours and many animals had died after convulsions as had been reported earlier. Others had been prostrate or lethargic but had later recovered. All in all, these results indicated that in populations of increasing size in a fixed space, but with adequate food and water, many animals succumbed to stress but the exact stressors could not be identified. Interestingly enough, in the largest populations the results were not so severe, possibly because animals could escape more easily from aggressive ones.

Barnett (1958) later investigated social stress in rats (*Rattus norvegicus* Berkenhout). He kept them in all male groups in large cages or in groups with both males and females in large cages, or as male–female pairs in small cages as controls. Strange males which he called

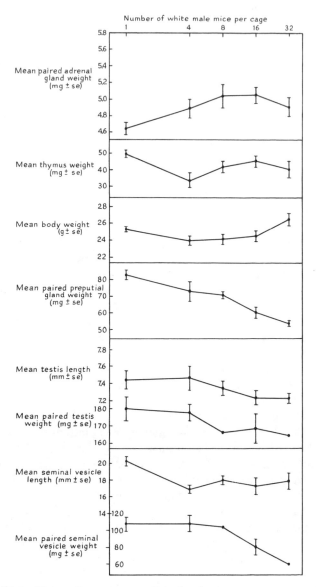

Fig. 11.4 Means of cage mean values for the organ measurements from albino male mice plotted against the logarithm of the population size. The standard error of the mean is given for the values from the albino mice. (From Christian, 1955.)

Fig. 11.5 An 'interloper' collapses during attack by a resident male. The resident still shows pilo-erection but ceases to pay attention to the interloper while it is immobile. (From Barnett, 1958.)

'interlopers' were introduced singly into the two types of colonies. The control males and those in all-male groups survived well but there was a high death rate in the male–female colonies and most interlopers died. Again there were the symptoms of convulsions, prostration and lethargy (Fig. 11.5). Deaths could not be attributed to wounding during fighting. The adrenal gland weights of the interlopers and males in the male–female colonies were significantly higher than those of the controls while those of males in all-male colonies did not differ significantly from the controls. However increased adrenal gland weight was not confined to males defeated in combat, for dominant males also had large adrenals.

From such very striking results particularly with rodents many writers have extrapolated the findings to other species but good corroborative evidence is generally lacking, particularly with regard to farm livestock (Wood-Gush *et al.*, 1975). A noticeable exception has been the work of Barnett and his colleagues on the sow (Barnett *et al.*, 1981). Their findings indicate that individual penning may be more disturbing to sows (as measured by their adrenal cortical responses) than keeping them in groups. Their results emphasize that social stress must be considered when housing and mixing animals.

POINTS FOR DISCUSSION

1. Single out the possible short-term stressors encountered by (a) the dairy cow, (b) the early-weaned piglet and (c) the hen in a battery cage with three other hens.
2. Discuss the effects of a hypothetical long-term stressor on any one type of livestock.
3. Try to name the stressors encountered by calves in transit on a long journey.

FURTHER READING

Barnett, J.L., Cronin, G.M. and Winfield, C.G. (1981), The effects of individual and group penning of pigs on total and free plasma corticosteroids and the maximum corticosteroid binding capacity, *Gen. and Comp. Endocrin.*, **44**, 219–225.

Beuving, G. and Vonder, G.M.A. (1978b), Effect of stressing factors on corticosterone levels in the plasma of laying hens, *Gen. and Comp. Endocrin.*, **35**, 153–159.

Friend, T.H., Polan, C.E., Gwazdauskas, F.C. and Heald, C.W. (1977), Adrenal glucocorticoid response to exogenous adrenocorticotropin mediated by density and social disruption in lactating cows, *J. Dairy Sci.*, **60**, 1958–1963.

Johnston, J.D. and Buckland, R.B. (1976), Response of male Holstein calves from seven sires to four management stresses as measured by plasma corticoid levels, *Can. J. Anim. Sci.*, **56**, 727–732.

Kilgour, R. and De Langen, H. (1970), Stress in sheep resulting from management practices, *NZ Soc. Anim. Prod.*, **30**, 65–76.

Shaw, K.S. and Nichols, R.E. (1964), Plasma 17- Hydroxycorticosteroids in calves – the effects of shipping. *Am. J. Vet. Res.*, **25**, 252–253.

Siegal, H.S. and Siegal, P.B. (1961), The relationship of social competition with endocrine weights and activity in male chickens, *Anim. Behav.*, **9**, 151–158.

Wood-Gush, D.G.M., Duncan, I.J.H. and Fraser, D. (1975), Social stress and welfare problems in agricultural animals, in *The Behaviour of Domestic Animals*, 3rd edn (Hafez, E.S.E., ed.). Bailliere Tindall, London.

12
EVOLUTION

It may seem rather odd to have to consider the evolution of behaviour patterns when what one is really interested in is animal production or veterinary practice. However, the essence of good stockmanship is in understanding the behaviour of one's animals. A consideration of the evolution of behaviour will not only show the processes that have been involved in domestication but will give a clearer insight into the changes that artificial selection can produce in behavioural traits. Finally it is only by considering the evolution of behaviour that we can understand some puzzling behaviour patterns.

The basic unit of behaviour with which we are concerned is the Fixed Action Pattern. These genetically determined behaviour patterns are characteristic for any species. They cannot be broken down into successive responses dependent on different external stimuli nor can their component parts change their relationships; only their degree of completeness can change (Hinde, 1970). When we consider the inheritance of behavioural traits we may in some cases be considering a single Fixed Action Pattern but often we will be considering the inheritance of a group of Fixed Action Patterns. The importance of displays in communication and courtship was discussed earlier and it is to this class of behaviour patterns that we will now turn our attention. Some displays are a single Fixed Action Pattern, others are a combination. Here we shall consider the origins and inheritance of some displays for they give us an important insight into the evolution of behaviour.

EVOLUTIONARY SOURCES OF DISPLAYS

The study of many displays shows that they have been derived from displacement activities and other behaviour patterns seen in conflict. In order for any of these behaviour patterns to become an effective display it must be obvious and if it is to become a signal, it must be standardized. Preening is a common displacement activity in birds. An examination of the mock preening display in mallards and related species illustrates some of the changes that have taken place in the course of evolution. In the mallard this display resembles normal preening very closely: the beak is drawn along the underside of the partly lifted wing producing a loud *Rrr* sound and as the wing is lifted the blue speculum is revealed. In the mandarin duck (*Aix galericulata*) this effect is enhanced by exaggerating the wing movement, raising it like a sail to show off the large rust-coloured primary feathers. However in this species the preening movements of the beak are reduced to the touching of a bright orange secondary feather. On the other hand the Shelduck has exaggerated the sound effects: by moving its beak in a rapid powerful stroke along the shafts of the wing quills it produces a low rumbling sound. In this example from these species the display has been changed by exaggeration and simplification of effective elements.

FORMATION OF DISPLAYS

In order for such conflict responses to become good displays they must be reliable so that the intentions of the animal are clear to the recipient. This means that the form must be invariable and that any temporal repetition must follow a set pattern, referred to as Typical Intensity. In the course of evolution additional morphological features such as conspicuous colouring of exposed parts may be added to ensure that the signal is seen or a vocal component is added. Which of these types of additions will occur, will depend on the habitat of the species and other factors. The process that ensures these changes is called 'ritualization'. A second process in which the display becomes independent of the original motivating forces so that the conflicting tendencies (e.g. fear and aggression) are no longer responsible for its elicitation is called 'emancipation'. When this has occurred the display is now motivated by the third motivational system in which it now plays a rôle. The exact nature of these two processes, ritualization and emancipation is not known.

Other conflict behaviour patterns that appear to be sources of new displays are intention movements, redirected activities, ambivalent postures and the overt elements of the autonomic responses found in conflict. Inhibition of movement due to conflict between staying and escaping may be the origin of the feigning death displays. Many threat postures appear to be derived from conflict between attack and escape and this hypothesis seems to fit many of the threat displays of species, particularly of fish and birds. For example, the waltz of the domestic cock is used not only as a threat display but also in courtship, particularly of strange hens in which case the male may be more aggressive than usual in order to assert his dominance. Analysis of this posture would indicate that it contains elements of aggression and fear. The outer wing of the cock is lowered as if he were chasing an inferior bird while the inner one is kept folded in a non-aggressive manner and the path traced never brings the waltzer face-to-face with the other bird, for a face-to-face stance is extremely aggressive, leading inevitably to an attack. Furthermore the neck is sometimes slightly retracted also indicating the presence of some fear. Nevertheless a waltzing cock is in a fairly aggressive state. In one study, 85% of waltzes were followed by some form of attack by the waltzer (Wood-Gush, 1956).

Other threat displays originate from the physiological side effects of conflict as a result of changes in the autonomic nervous system. For example the erection of hair in certain areas of the body has come to be a noticeable factor in the threat displays of some species. The threat displays of ungulates (see Chapter 2; Walther, 1974) have not evolved from conflict between attack and escape but are intention movements to attack, although some may be derived from defensive behaviour in certain species in which attack and defence involve different behaviour patterns. However, from careful consideration of Walther's writings it would seem that fear cannot be ruled out in the evolution of some displays at least. Redirected aggression towards some environmental object may be seen in some ungulates but the communicative value of these behaviour patterns is unknown, while those displays evolved from intention movements are of highly communicative value (Walther, 1974). He also suggests that threat displays have another function in that they allow one opponent to become aware of the other's aggressive intentions and to avoid losing in a chance attack.

Relatively few studies have been performed in which the genetic basis of Fixed Action Patterns have been tested in vertebrates. In such a study, Sharpe and Johnsgard (1966) investigated the displays of 23

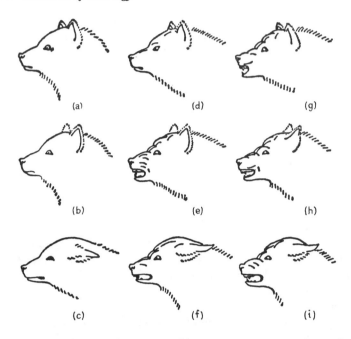

Fig. 12.1   Facial expressions of the dog that result from a superposition of various intensities of fighting and flight intentions. (a–c) Increasing readiness to flee. (a–g) Increasing aggression. (From Lorenz, 1981.)

F_2 drake hybrids from P_1 crosses involving mallard ducks and pintail drakes. Pair-formation displays shown by the hybrids resembled those of the parental species in those cases in which the display is very similar in both species. In other pair-formation displays they tended to be intermediate with some components like one species and others like the other. For example, one display sequence found in both species is 'head-up-tail-up' and 'nod swim', but there are differences in the form of the displays. During the 'head-up-tail-up', the head is raised while the tail is strongly lifted. The bill is then pointed to a specific female and a call is uttered. The mallard performs this pointing while the tail is still being raised whereas the pintail lowers it just before pointing (Fig. 12.2). The mallard drake remains in this pointing posture for a short time whereas the pintail drake reorientates the body into a straight axis with the head and holds this position momentarily. This is called 'facing'. In the hybrids tail-

Fig. 12.2   The 'head-up-tail-up' display in (a) the mallard drake and (b) the pintail drake. (From Lorenz, 1971.)

lowering and bill-pointing followed the mallard pattern closely in three cases and the pintail pattern in eight. Facing in the pintail fashion was done by nine drakes and in the mallard fashion by two.

In the case of displays performed by only one of the parental species these were performed either completely or not at all or, in the case of some displays, in a rudimentary or aberrant manner. Furthermore the sequences of some displays was altered in some hybrids; a finding which had been reported in F_1 hybrids involving mallard × black duck (*Anas rubripes*) and mallard × Florida duck (*A. fulvigula*) crosses (Ramsay, 1961).

In the mallard × pintail study the plumage of the hybrids varied from mainly mallard-type to mainly pintail-type with the plumage type being highly correlated with the type of display. In other words, birds with mallard-like plumage tended to have mallard-like displays and vice versa. In another study Hinde (1956) investigated the displays of three species of finches, the goldfinch, the greenfinch and the canary (*Carduelis carduelis, Chloris chloris* and *Serinus* sp. respectively) and the F_1 hybrids. Unfortunately the crosses were not reciprocal and the hybrids' sterility prevented the breeding of an F_2 generation but the results bear out the inferences made earlier; that the homologous displays in closely related species are Fixed Action Patterns that are heritable like morphological traits.

In these species there is a threat display the 'head forward threat' in which there are slight differences in the three species depending on the amount of threat. In the goldfinch, wing-raising is more marked and sometimes the wing is waved up and down. Fluffing of the feathers during threat is more common in this species than in the others. Flicking of the wings, on the other hand, is more common in greenfinches. These sort of differences were found to be inherited. For example, in crosses involving goldfinches, the hybrids fluffed their feathers, but to a lesser extent than the pure goldfinch. Another threat posture, seen only in the greenfinch, in which the neck is stretched and elevated and the head jerked upwards is called 'head up' and appears to be dominant.

In the greenfinch and the canary the male courtships are similar, but in a display called 'sleeked – wings-raised' the neck is raised more in the greenfinch. In these hybrids the neck position was found to be intermediate, while in crosses between greenfinches and goldfinches this display was similar to the greenfinch display. In the goldfinch the male has a striking pivot display in courtship in which he swings his body laterally in a stereotyped way. In the other two species the movement does not appear to have been ritualized and it appears as a series of intention movements. In the goldfinch × greenfinch hybrids the display was intermediate between the stereotyped movements of the goldfinch and the fortuitous intention movements of the greenfinch.

Finally in Hinde's study there is evidence for the inheritance of a displacement activity, which as has been mentioned, is one of the conflict activities thought to be the main sources of new displays. The goldfinch cock during courtship shows incomplete displacement breast preening which is less common in the other two species, but which was frequently seen in hybrids from crosses involving goldfinches. Thus one of the hypothetical potential building blocks for new behaviour patterns has been shown to be a heritable trait.

As mentioned above hybrid drakes were studied by Ramsay (1961). Again they were F_1 hybrids only, and the results can only be accepted tentatively, but the point of interest is that the order of the displays was different in the hybrids from either of the parent species. If accepted as true genetic effects, these findings suggest that changes in the sequences of behaviour may also be potential sources of new displays.

There is little evidence for the genetic basis of other conflict activities such as 'redirected activities' which, as postulated, could also be sources for new displays. Studies on different strains or inbred lines can sometimes suggest that certain sorts of conflict activities have a genetic basis. For example, the two strains of domestic hen having different behaviour patterns in battery cages in the hour or so before laying (mentioned earlier) show that there are intra-specific differences in response to thwarting. Further studies on these strains have shown that these differences are genetic (Mills and Wood-Gush, 1982). However the detailed genetic analysis is still in progress. This investigation fills an important gap in our knowledge about the derivation of displays. If they are indeed derived from conflict behaviour then it is important that such behaviour should have a genetic basis on which natural selection may act. Up to now very little attention has been paid to this aspect.

BEHAVIOURAL ADAPTATION

A central theme in the theory of evolution is adaptation; species are considered to be generally adapted to their ecological niches. Detailed observations on a population in its natural habitat will often show up some of the more obvious adaptations. Some studies have gone further than this and have attempted to study the development and control of these species-specific behavioural adaptations. One such study was carried out by Tschanz and Hirsbrunner-Scharf (1975) on the behaviour of the guillemot chick which is hatched from a single egg clutch and reared on cliffs on narrow ledges occupied by many other breeding conspecifics. Detailed observations had shown that the chicks have about twelve behavioural traits that appear to be adaptive to this perilous habitat. They rest in a sitting position between the parent bird and the rock face. Sitting rather than lying prevents them from getting soiled on the ledge which is covered with droppings and the position against the rock face prevents them from being pushed off or being taken by predators. They are fed one fish at a time, which is carried and delivered by the parent bird in such a way the neighbouring birds cannot easily see it and steal it. The chick is highly attentive to the acceptance call of the parents and if it has left the breeding site it will return on hearing the call. The related razorbill also breeds on sea cliffs but uses caves and crevices for nesting and rearing the single young. However, razorbill pairs nest separately from other conspecifics and many of the guillemot chick adaptations are missing in the razorbill chick. They nest in a lying position, they move around more and they are offered several fish at a time since there is little chance of theft. Tschanz and Hirsbrunner-Scharf investigated the behaviour of guillemot and razorbill chicks hatched in the laboratory and reared under identical conditions. Several of the differences between the species seen in the wild were obvious in the laboratory. The guillemots tended to rest in a sitting position while the razorbill chicks lay with breast on the floor significantly more. The guillemot chicks chose the darkest area of their pen and stood near a wall significantly more than the razorbill chicks, thus suggesting a phylogenetic origin for these adaptive traits. The authors then carried out some cross-fostering on the cliffs. The razorbill chicks were found to be poorly adapted to living on the crowded, open ledges and to the method of feeding used by their guillemot foster parents. They became soiled and cold; they wandered about and did not respond to parental calls; they pecked at the fish being offered by the guillemot foster parent which was disturbed by the action, with the result that the fish was often stolen by a neighbour. Nearly half the razorbill chicks perished but as they got older so they and their foster parents adapted their respective behaviour patterns to a certain degree. The guillemot chicks also had some difficulties but these were minor and losses were few. In general, both types of chick showed behaviour specific to their species, but also showed that they were capable of modifying these to some extent. Such modification would also be adaptive, for it would allow a species to occupy new habitats and to meet new challenges.

COMPARISONS OF CLOSELY RELATED SPECIES

Behaviour patterns with a genetic basis will, like morphological and

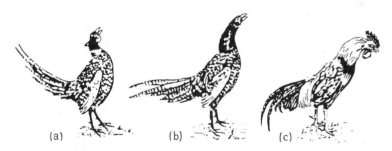

Fig. 12.3 The crowing posture of the ring-necked pheasant (a) a domestic cock (c) and a hybrid of both species (b). (From Immelmann, 1980; after Stadie, 1968.)

physiological traits, be subjected to natural selection. Furthermore, as we have seen, behavioural traits in closely related species will be more similar than in distantly related species. In fact behavioural traits have been used like morphological traits in taxonomy.

If we consider species that are judged by a number of criteria to be closely related we can see that not only do they have a number of behaviour patterns in common but that gradations in one or more behaviour patterns can be traced from one species to another changing in complexity, and possibly function, through several closely related species.

The further apart species are phylogenetically, the greater the behavioural differences will be. As Lorenz (1971) has shown with duck and geese, the more closely related the species are, the more displays they have in common. A few are shown in Fig. 12.4 to illustrate this point. The mallard, the spot-billed ducks and Meller's duck all belong to the same genus and share many behaviour patterns. (Fig. 12.4). The pintail and South American pintail, belonging to the same family as mallards but to a different genus, do not have as many behavioural traits in common with the mallard and its sibling species. The muscovy on the other hand, belongs not only to a different species and genus but to a different family and shares relatively few of their traits.

Similarly, Tinbergen (1959) compared two closely related species, the blackheaded gull (*Larus ridibundus*) and Hartlaub's gull. It will be recalled that analysis of displays revealed that many can be explained by the arousal of conflicting tendencies, namely fear, aggression and mating. In these two gull species the differences in the displays appear to be due to greater exaggeration of the fear reactions of the Hartlaub's gull. To take two examples: in Hartlaub's gull in one display, 'head flagging', there is a thinner neck and sleeker plumage; in the second display 'forward posture' the bill is held pointing upwards. These features are absent in the 'head flagging' and 'forward posture' displays of the blackheaded gull but are found in the escape reactions of both species. Comparisons like this can show us the sources of evolutionary changes which can eventually lead to greater divergence.

Another such study has been carried out on a group of species belonging to the parrot family. Dilger (1962) investigated the displays of eight species of the genus *Agapornis* which form a series that includes three primitive species and a group of five species and subspecies, which inhabit Africa from Ethiopia to Malawi. The

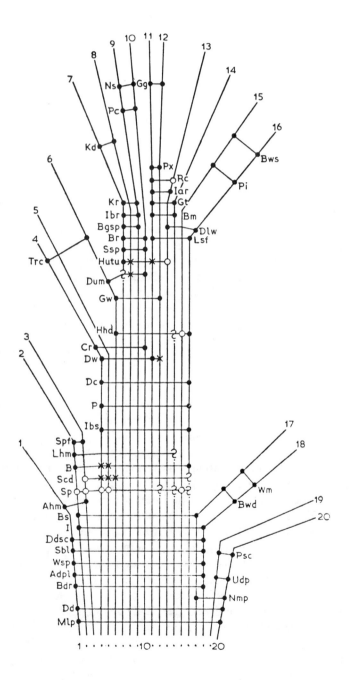

primitive species differ in a major behavioural aspect from the five advanced species and subspecies in that the social unit is the mated pair while the others are colonial. In the sexual behaviour of the species investigated, the male performs displacement scratching. In the three primitive species this consists of the male scratching his head with the foot nearest the female and it appears to be done when he is frustrated. In the more recently evolved species it has become a form of display. The scratching is more perfunctory and in the three subspecies it is directed towards the bill and either foot may be used. Furthermore while it is performed in the presence of the female, she does not appear to be frustrating him, thus it appears that the function may have changed.

The more recently evolved species have also evolved a form of display fighting in which there is bill fencing and sharp pecks are given at the opponent's toes but the fighting is never lethal. In the non-colonial primitive pairs there is no such ritualized fighting, and fighting, if it occurs, is lethal; instead they have evolved elaborate series of threat and appeasement postures.

In another display 'squeak-twittering', the male of the three primitive species utters a series of highly-pitched calls when the female thwarts him by disappearing into the nest cavity. The sounds are quite variable in pitch and purity of tone and there is no recognizable rhythm. In the other species the call is rhythmic, purer in tone and less variable in pitch. It is not confined to occasions when the female is entering the nest cavity but sometimes given in her presence. It therefore has changed in its physical characteristics and also possibly in the context in which it is given. Similar examples of displays in closely related species to these will be discussed when we come to domestication, and other studies have been carried out using a variety of species including primates.

Fig. 12.4 The vertical lines indicate the different species, the horizontal lines traits common to them and the letters the different traits. A cross indicates the absence of a character, a question mark uncertainty about its occurrence in a species and circle indicates that a trait has undergone emphasis and differentiation. 1, Muscovy duck; 10, mallard, spot-billed ducks and Meller's Duck; 11, South American pintail; 12, pintail. (From Lorenz, 1971.)

BEHAVIOUR AND SPECIATION

The process of speciation or the formation of new species is a facet of evolution in which behaviour can play a decisive role. In order for two populations to diverge to the extent that successful cross-breeding is impossible, it is necessary for the populations to be isolated from one another so that barriers to their crossing are formed. These barriers may be dependent on chromosomal or gene incompatabilities, so that only non-viable hybrids are produced. On the other hand the barriers may be ecological, geographical or behavioural so that, while hybrids may be produced in the laboratory, they are never produced in the wild because the two species are segregated ecologically or geographically, or their behaviour differs and thus forms a barrier. In general all these types of barriers probably operate together as isolating mechanisms.

Behaviour can help to isolate populations in a number of ways. Sexual imprinting will lead to animals mating with ones of a certain phenotype as we discussed earlier: Brown Leghorn and White Leghorn fowls will only tend to mate with their own breeds if they have no experience of the others (Lill and Wood-Gush, 1965). In their case this may be only a temporary incompatability, but permanent cases of sexual imprinting have been reported (Immelmann, 1975). Early experience can also effect other preferences besides sexual preferences, for both food and habitat preferences can be established early in life. These processes, which, as mentioned in Chapter 1, resemble imprinting, are highly resistant to change and would serve to isolate populations (Immelmann, 1980). In song birds, differences in song as we saw earlier are partly learnt in most species so far studied, so that birds acquire a local dialect. Such dialects may also act as isolating mechanisms. Although species probably also develop mechanisms that prevent a high degree of inbreeding, large differences in behaviour will act as isolating mechanisms.

SOME CONSIDERATIONS IN RELATION TO DOMESTICATED ANIMALS

In this chapter we have discussed Fixed Action Patterns and shown that in evolution they behave like morphological traits and remain species-specific. They do not change except in their degree of completeness. We saw too that species-specific behaviour in the case of the razorbill chicks can be also somewhat modified by experience. However this does not mean that the Fixed Action Patterns themselves were modified but rather that the order of a number of such patterns was changed through learning. We also saw, in the case of the breeding experiments involving duck hybrids, that some traits, which are either Fixed Action Patterns or chains of them, appear to be dominant. Others are recessive and sometimes new combinations can arise by cross-breeding. Some of these points will be encountered when we consider domestication which is essentially a special type of evolution.

POINTS FOR DISCUSSION

1. Discuss the differences between speciation and the development of new breeds.
2. When a dog bites the postman delivering mail to the house, the behaviour may be considered to be due to the territorial behaviour of the dog. Discuss any other behavioural traits in our domestic and domesticated animals that are a nuisance now, but which were advantageous to the species in their evolutionary history.
3. Discuss Emancipation and Ritualization.

FURTHER READING

Geist, V. (1971), *Mountain sheep. A study in behaviour and evolution*, Univ. Chicago Press, London.

Hinde, R.A. (1970), *Animal Behaviour. A synthesis of ethology and comparative psychology*, 2nd edn. McGraw-Hill, Kogakusha Ltd, London, pp. 657–689.

Kiley-Worthington, M. (1976), The tail movements of ungulates, canids and felids with particular reference to their causation and function as displays, *Behaviour*, **56**, 69–115.

Manning, A. (1979), *An Introduction to Animal Behaviour*. Edward Arnold, London, pp. 210–229.

Vines, G. (1981), Wolves in dogs' clothing, *New Scientist*, **91**, 648–52.

Wood-Gush, D.G.M. (1975), Nest construction by the domestic hen: Some comparative and physiological considerations, in *Neural and endocrine aspects of behaviour in birds* (Wright, P., Caryl, P.G. and Vowles, D.M., eds). Elsevier, Amsterdam, pp. 35–49.

13

DOMESTICATION

Domestication has been defined by Rattner and Boice (1975) as the removal of organisms from some natural selective pressures over generations. However this definition is not entirely satisfactory for it could as easily apply to certain zoo animals as to animals that are conventionally considered to be domesticated. The latter fulfil a variety of functions for man and it is the conscious genetic selection for adaptation to these functions that today largely distinguishes domestic animals from the inmates of a modern zoo, which are also being protected from some pressures of natural selection that would operate in their natural habitats. Additionally most domestic animals are more docile than most zoo animals. However domestication must be distinguished from tameness, which is defined as the elimination of the tendency to flee in the presence of man. Tameness may be learnt through early association with man, e.g. imprinting of young animals on to man, or it might occur through lack of contact with man. Several species of animals on the Galapagos Islands do not flee from man, probably due to the fact that they have not been hunted. Furthermore, domestication can involve many different sorts of relationships with man, from the intimate relationship between a pet dog and its owner to the loose association found, say, on an ostrich farm or extensive sheep farm. In either event the species concerned is able to fulfil the goal desired by the owner and to breed successfully in the environment supplied by man. In some cases reproductive success is now completely dependent on man's intervention, as in the case of the bulldog, the non-broody strains of fowl, and some broad breasted strains of turkeys in which the males are unable to copulate successfully.

BEHAVIOURAL FACTORS FAVOURING DOMESTICATION

Hale (1962) listed the behavioural traits which would enhance the process of domestication.

1. Group structure
   (a) Large social group
   (b) Hierarchical group structure
   (c) Males affiliated with female group
2. Sexual behaviour
   (a) Promiscuous matings
   (b) Males dominant over females
   (c) Sexual signals provided by movements or postures. (Signals provided by colour markings, etc., would be disadvantageous.)
3. Parent–young interactions
   (a) Critical period in development of species bond, i.e. learning is involved and the bonding is not established mainly by innate mechanisms
   (b) Female accepts other young soon after parturition or hatching
   (c) Precocial development of young
4. Responses to man
   (a) Short flight distance to man
   (b) Least disturbed by human presence and activities
5. Other behavioural characteristics
   (a) Catholic dietary habits
   (b) Adaptability to a wide variety of environmental conditions
   (c) Limited agility.

THE ORIGINS OF DOMESTICATION

Most of the main agricultural species were probably domesticated in Neolithic times in certain areas. In South East Asia domestication of the fowl, duck, goose, dog and pig occurred, while South America saw the domestication of llamas, alpacas, guinea-pigs and muscovy ducks, and Central America, the turkey. A major region however was the Middle East where goats, sheep, cattle and pigs were involved (Fig. 13.1).

It has been suggested (Sauer, 1952) that our domestic agricultural animals were not originally domesticated for their food potential but as pets and for enjoyment. The domestic fowl, he suggests, was first used for cock-fighting and only later primarily as a source of food. Some, like the dog, probably started as a scavenger with pups being kept as pets. These ideas are based on observations that many primitive people tame and rear wild animals. However, the food potential of our main agricultural animals has been used for a long time and by the Roman period there was, in Italy, an organized poultry industry with specialized breeds and intensive husbandry. The record of the horse as a working animal is even older: they have been used for transport for at least 5000 years (Zeuner, 1963).

THE PROCESS OF DOMESTICATION

Domestication may be considered as a special form of evolution and in the generations before man started applying conscious genetic selection to his animals, there was unconscious selection for a number of appropriate traits. It is possible to see how the behaviour of some species has been modified by this genetic manipulation by comparing the behaviour of the domestic species with that of the putative ancestor or closely related species. Such a method of course cannot be fool-proof, for not only is the ancestry often doubtful but the behaviour of the ancestral species may have undergone modification since the separation of the two forms. Further insight into the process can be gained by looking at the behaviour of all common domesticated species and trying to sift out any common trends.

Two species will now be considered in detail: the domestic duck and the domestic dog.

The domestic duck

Desforges and Wood-Gush (1975a, 1975b, 1976) compared the behaviour of Aylesbury ducks with that of mallards which are thought to be the ancestral species. Hybridization is common between mallard and domestic duck, and pains were taken to get a mallard population that was likely to be uncontaminated by genes of the domestic duck. For this purpose eggs were obtained from a nature reserve with a high proportion of migratory duck. The mallards, which were each handled for a few minutes daily for their first three months after hatching, were treated in exactly the same way as the Aylesburys from hatching. They were kept under semi-intensive conditions in pens, in small mixed groups of males and females. In addition, some

Aylesburys were kept on a large out-door pond on a farm, which resembled many of the habitats taken up by the mallard, to see how their behaviour resembled that of the mallard under such conditions.

The sexual displays of mallards which have been extensively reported in the wild can be divided into three major categories; (a) social courtship displays, (b) pair-formation displays and (c) copulatory displays. Under the conditions of the observation pens, the mallards were found to show the normal range of displays, and those done usually while swimming were done either standing or moving about the pen. Qualitatively the courtship displays of the two types of duck were very similar, but there were quantitative changes in the frequencies at which some displays were given and in some the intensity of the display (completeness and frequency) was lower in the Aylesburys. In some others the vocal component was missing in the Aylesburys, so that a feature that drew attention to the display was lost.

A quantitative comparison of three frequent social displays of the drakes is given in Table 13.1.

In the 'grunt-whistle', the drake touches the water with the tip of the bill, then rears up, raking the bill through the water throwing up small drops of water. Before the body resumes the normal swimming position a loud clear whistle is given followed by a faint grunt. In the 'head-up-tail-up' the drake throws the head upwards and backwards while spreading and vertically erecting the tail feathers. The wings are slightly raised and the speculum exposed (see Chapter 12) In the 'down-up' the drake dips the breast deep into the water then raises the head high whistling and calling. In all these the Aylesbury has lost the whistle and shows less intensity in the displays.

In the pair-formation displays a common one by the female is inciting. When doing this the female selects a certain drake, goes

Table 13.1   A comparison of three common social displays

Type	No. of drakes	Grunt-whistle*	Head-up-tail-up*	Down-up*
Mallard	6	42	49	55
Aylesburys	6	41	34	7

* All these displays have vocal components in mallards but none was heard in the Aylesburys. (After Desforges and Wood-Gush, 1976.)

towards him repeatedly moving her head backwards over one shoulder and forward again keeping her bill pointed down and giving a special call. Once pair-formation has taken place, the female uses this display to stimulate her mate to attack other drakes. The mallard females in this study incited to a single male whereas the Aylesbury females each incited several males to mate.

A drake display belonging to the pair-formation group is 'preen-behind-the-wing'. It is common in paired mallards in the wild, but was rarely seen in the Aylesburys on the pond; another pair-formation display, ritualized drinking, in which the pair drink together in long and intensive bouts, was common in the mallard but absent in the Aylesburys.

In the third category of sexual behaviour, copulatory behaviour, there was no difference between the two types with respect to the display immediately preceding copulation nor in the motor movements of copulation, but the Aylesbury ducks copulated with several drakes whilst the mallards copulated with only one other member of the group with which they had formed a pair bond. Rape (copulation without the preceding display of pre-copulatory pumping) was common in the Aylesburys but rare in the mallards under these conditions.

In summary then we can see that the form of the displays is the same in both types although one pair formation display was not seen in the Aylesbury. The main changes seem to be in the loss of the vocal component in some and a drop in intensity. The mallards formed pairs in which members of a pair fed together and slept close together but there was no evidence of this in the Aylesburys, either in the pens or on the pond. Although the Aylesbury ducks tended towards promiscuity they showed some preference for certain males. Furthermore, while some mallard drakes form territories under these conditions, no Aylesbury males did so, even on the pond.

In general the mallards were much more aggressive towards one another than the Aylesburys which were far more tolerant towards birds feeding next to them and also rested closer to pen-mates than the mallards.

The Aylesburys also habituated to novel objects much more quickly than mallards. In one experiment a stuffed hen was placed in the home pens of the two types, and the model was so placed that the ducks had to pass it in order to reach their food and water. After 45 minutes the model was removed. For six days this procedure was repeated and in

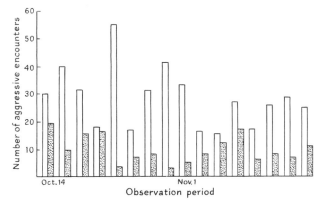

Fig. 13.1 A comparison of the frequency of aggressive encounters in equal sized groups of mallards □ and domestic ▨ ducks, similarly kept but with the domestic ducks having slightly less space per bird (From Desforges, 1973.)

each trial the number of crossings of the pen by the birds was noted. During the first trial none of the Aylesburys crossed the pen, but during the second trial all did so at least once. In later trials there was a tendency for the females to cross much more often than the drakes. The mallards however never crossed the pen during any of the trials. In another experiment the reactions of the birds to a human being were tested by releasing a duck in a runway and measuring the time the bird took to reach the far end of the runway. The mallards moved away not only nearly twice as fast as the Aylesburys, but the latter often stopped en route and showed little immediacy to move. When presented with novel food in the form of dyed food pellets, the Aylesburys took longer to eat this than normal food, but accepted it whereas the mallards did not eat the novel food in any of the five trials of the experiment.

From these and other experiments from this study it may be concluded:

1. The motor patterns of the display and consummatory behaviour of the two types don't appear to be different qualitatively.
2. The main differences in displays are in intensity, loss of a component or loss of a display in the domestic form.
3. In the domestic duck there is more promiscuity and a loss of pair formation and territoriality.

Table 13.2 Aylesbury and mallard ducks: average latency(s) to feed after food deprivation lasting 6 h. (From Desforges and Wood-Gush, 1975b.)

Trial no.	Normal food in normal feeder	Normal food in glass bowl	Coloured food in glass bowl
Aylesbury			
1	112	64	557
2	61	35	157
3	27	189	205
4	41	49	149
Mallard			
1	263	1397	No eating
2	441	1281	No eating
3	196	1157	No eating
4	282	907	No eating
5	140	946	No eating

4. There is less intra-specific aggression in the domestic duck and more tolerance towards conspecifics.
5. There is no seasonal restriction to sexual behaviour in the domestic duck and it appears earlier than in the mallard.
6. The domestic form habituates to novel stimuli more quickly and is less disturbed by the presence of man.

The domestic dog

Its ancestry is not exactly known, for although descended from the wolf, hybridization with the coyote and jackal may have taken place (Fox and Bekoff, 1975). It is therefore difficult to make a strict comparison between the dog and its direct ancestor (assuming that it is the wolf) and to attribute all differences as being due to domestication.

Although man has bred a large number of varieties of dog (possibly as many as 800), the behaviour patterns have not been altered basically. What has happened is that certain traits have been enhanced in some breeds and others suppressed. Scott and Fuller, and their co-workers, have made a comparison of 50 main behaviour patterns in dogs and wolves (Scott, 1950). They watched the behaviour of Scotch Terriers, Fox Terriers and Beagles kept in fields of one acre per litter. This relatively spacious and free environment allowed many behaviour patterns to be shown which would have been dormant in

Table 13.3 Expression associated with an increase of social space and aggression. +, frequent; ±, infrequent; −, not observed (i.e. when at very close range). (From Fox, 1971.)

	Coyote	Wolf	Dog
Head high-neck arched	+	+	+
Gape	±	−	−
Growl	+	+	+
Bark	±	±	+
Agonistic 'pucker' (horizontal contraction of lips)	+	+	+
Agonistic baring of teeth (vertical contraction of lips)	+	+	+
Snapping of teeth	±	+	+
Ears erect and forward	+	+	+
Direct stare	+	+	+
Eyes large	+	+	+
Ears flattened and turned back	+	+	+

more restricted environments. The animals were also allowed to breed and so the full cycle of behaviour could be watched. It does not seem from this study that any new behaviour patterns have come into being, but the workers emphasize that some breeds of dogs have been selected for prolongation of patterns of behaviour which originally were typical of young puppies. For example, attention-seeking behaviour such as whining may be extended far into life and the same is true of play behaviour (e.g. playful fighting).

Fox has made a detailed study of the agonistic displays of canids closely related to the dog and his findings again emphasize the point that domestication has not altered the behaviour patterns basically, as shown in Tables 13.3 and 13.4. Furthermore in the wolf, coyote and dog the same body areas are preferentially attacked both in play-fighting and ritualized aggression, namely the scruff of the neck, the throat and cheek.

Breeding and training by man have not changed the expressions, but the probability of whether they appear or not, e.g. guard dogs are said to attack without giving any facial expression that shows the intention to do so, and in some other dogs the ears have become too

Table 13.4   Components of expressions associated with decrease of social distance and submission. (From Fox, 1971.)

	Coyote	Wolf	Dog
Head lowered and neck extended horizontally (crouch)	+	+	+
Ears flattened and turned down at the sides	+	+	+
Submissive grin (horizontal retraction of lips)	+	+	+
Play face	+	+	+
Licking cut off	+	+	+
Licking social greeting	+	+	+
Licking (intention)	+	+	+
Nibbling	±	+	+
Jaw wrestling (play)	+	+	±
Whining and whimpering	+	+	+
Looking away	+	+	+

pendulous to be used as signals, or the face so covered with hair that the signals cannot be seen. The direct stare is very important and in a wolf pack a direct eye contact with the alpha male, even if 20 yards away, may cause the subordinate animal to go into a submissive posture. Similarly eye contact with a dog in its territory may result in it attacking one. Looking away by turning the head is, as we have seen, a common posture in submissive situations in all three species and in dogs occurs frequently when they are in conflict (e.g. approach-avoidance situations). It is part of a common stereotypy in Daubermann Pinchers (Fox, 1969) in which the dog turns its head and chews its flank. It is thought to be caused by anxiety.

The position of the tail in displays of dominance differs between the dog in which it is curled upwards, and the wolf and coyote, in which it is held straight out.

Barking is a good example of a behaviour pattern in which the threshold has been lowered by domestication to such an extent that it is released by a bewildering variety of stimuli in some dogs. In the other two species it would appear to have other functions; it may infrequently occur in agonistic encounters and also in connection with howling which serves to maintain contact between members of a wolf pack. Possibly barking in dogs has been emancipated by domestication.

Many scent marking traits are the same in the three species; the animals will roll in certain odours (Fox, 1971) and will urinate on strange objects. In the territories of wolf packs are scent posts which are marked with urine and if one is marked with strange urine the pack become very excited and members of the pack urinate repeatedly on that scent post. In the Canidae the peri-anal glands release an odour during defaecation which may impart an individual identity to the odour of the faeces and so faeces, like urine, may be used for territorial marking.

Many components of sexual behaviour are the same in the three species, but the bitch comes into oestrus twice a year whereas the female coyote and wolf do so only once a year. However, one breed of dogs, the Basenji, is an exception in that the bitch comes into season once a year. The coyotes tend to form pair bonds that are very long lasting and the wolves too form pair bonds, while the bitch is promiscuous. Thus it appears that, as in the case of the Aylesbury duck, the domesticated form is not only more fecund but also more promiscuous. Parturition in dogs resembles that of the other two species in almost all the main features (Fox and Bekoff, 1975). Regurgitation to feed the young is found in the wolf and coyote, but only occasionally in the domesticated dog and it serves as an example of a behaviour pattern in which the threshold has been raised during domestication

The social structure of wolves, coyotes and dingos differ very much. The wolf has an intricate pack system; the coyotes live in pairs on territories and may hunt with their semi-adult litter, whereas the dingo may hunt alone, or in pairs or in groups that appear to be aggregations at a food source rather than the cohesive pack of the wolf. In some breeds of dog there has been selection for hunting in packs, but these are more like the aggregations of the dingo than a true pack. In other breeds such as the terriers this tendency has been selected against.

From these two case histories and from studies on other domestic species it appears that the process of domestication by conscious or unconscious selection has involved:

1. Changes in threshold at which many behaviour patterns occur. The threshold may be so high in some cases as to make the occurrence a very rare event. Often, however, the behaviour pattern may be different in intensity and completeness. With a drop in threshold the behaviour may become more frequent, as in the case of barking in the dog or in the mating behaviour of Aylesbury ducks.
2. Changes in responses to key stimuli: the animals may respond to new stimuli. At one level this may involve the relatively rapid acceptance of novel food as in the case of the domestic ducks, or, more fundamentally, in the case of the dog whose social attachment to man can be explained in terms of the dog reacting to a pack leader or dominant pack member. Some of these changes will be largely determined by learning and others not. Sometimes the response may be given to a reduced number of stimuli, e.g. the domestic duck will respond to a displaying drake although he no longer has the specific markings which are emphasized by the display of the mallard drake. In other cases the domesticated stock may habituate to stimuli that affect the wild species.
3. Infantile behaviour patterns have been prolonged in some cases, e.g. in the dog play endures for much longer than in other Canids.
4. It is possible that certain types of learning have been enhanced by domestication but this is difficult to prove.

The changes occurring in the thresholds of response may involve changes in the CNS. For example, Phillips and van Tienhoven (1960) found wild ducks which did not lay in captivity came into reproductive condition after lesions were placed in the archistriatum. Other cases may involve the endocrine system directly, e.g. the speed and rate of adrenalin release.

So far there is little evidence of new behavioural patterns in domestic species and such changes that have been studied can be explained in the terms discussed above.

POINTS FOR DISCUSSION

1. Discuss the difference between domestication and tameness.
2. Attempts are being made to domesticate the Red deer. Discuss some of the difficulties that you feel may be encountered.
3. Discuss the physiological changes that may be involved in domestication.
4. Some breeds of dog are 'highly-strung' due to inbreeding. Discuss

this statement in relation to the process of domestication.
5. Compare the behaviour of one species of farm livestock kept intensively or extensively with a feral population. What do you think can be gained from such studies?

FURTHER READING

Experimental studies on the process of domestication (additional to those mentioned in the text).

Boice, R. (1977), Burrows of wild and albino rats: Effects of domestication, out-door raising, age, experience and maternal state, *J. Comp. Physiol. Psychol.*, **91**, 649–661.

Connor, J.L. (1975), Genetic mechanisms controlling the domestication of a wild house mouse population (*Mus. Musculus* L.) *J. Comp. Physiol. Psychol.*, **89**, 118–130.

Fox, M.W. (1976), Effects of domestication on prey catching and killing in beagles, coyotes and F_2 hybrids, *App. Anim. Ethol.*, **2**, 123–140.

Leopold, A.S. (1944), The nature of heritable wildness in Turkeys, *Condor*, **46**, 133–197.

Vines, G. (1981), Wolves in dog's clothing, *New Scientist*, **91**, 648–652.

Domestic and feral populations in the wild.

Cattle
Hall, S.J.G. (1979), Studying the Chillingham wild cattle, *Ark*, **6**, 72–79.

Domestic Fowl
McBride, G., Parer, I.P. and Foenander, F. (1969), The social organisation of feral domestic fowl, *Anim. Behav. Monogr.*, **2**, 125–181.

Wood-Gush, D.G.M., Duncan, I.J.H. and Savory, C.J. (1978), Observations on the social behaviour of domestic fowl in the wild, *Biol. Behav.*, **3**, 193–205.

Horses
Feist, J.D. and McCullough, D.R. (1975), Reproduction in feral horses, *J. Reprod. Fert. Suppl.*, **23**, 13–18.

Feist, J.D. and McCullough, D.R. (1976), Behaviour patterns and communication in feral horses, *Z. Tierpsychol.*, **41**, 337–371.

Tyler, S.J. (1972), The behaviour and social organisation of the New Forest Ponies, *Anim. Behav. Monogr.*, **5**, 85–196.

Sheep
Grubb, P. and Jewell, P.A. (1966), Social grouping and home range in Feral Soay sheep, *Symp. Zool. Soc. Lond.*, **18**, 179–210.

14
ANIMAL WELFARE

Animal welfare has been defined as being a state of complete mental and physical health in which the animal is in harmony with its environment (Hughes, 1976a). Although this should be our aim, it seems impossible, for we will never know when an animal is in harmony with its environment or when it has complete mental health. To define animal welfare from a practical point of view is a much more difficult thing. A number of possible criteria have been discussed and we shall review them .

THE ASSESSMENT OF WELFARE

Production records

These are much favoured by the various sectors of the livestock industry as an indicator of well-being but, as Dawkins (1976) points out, this criterion was dismissed by the Brambell Committee on the grounds that good growth rate and condition may not be incompatible with periods of acute but transitory physical or mental suffering. For example, it is quite possible for a veal calf or broiler that has become lame through unsatisfactory housing to give an economic return. Furthermore, if Selye's postulation of a triphasic response to a stressor is considered (Selye, 1960), then it is easy to visualize that animals, even if continually under stress, may give an economic return if they are kept for relatively short periods before being sold. In effect they are marketed when they are in Selye's 'stage of resistance'. Furthermore, as production levels are not static what level of production should be used? In some enterprises low levels of production are economic because of low capital input. These need not necessarily be enterprises with extensive systems of husbandry, but intensive systems with a poor level of husbandry and in which economies in husbandry, housing and feeding allow a lower production level. McBride (1968) has pointed out another difficulty; in many enterprises the husbandry is based on average requirements for a class of animals and many individuals are kept under conditions which will not ensure that they can reach their full potential.

Choice of environment

The adequacy of the environment supplied by modern systems is one of the crucial points in animal welfare. Therefore allowing the animal to choose its own environment has been advocated by a number of workers, and a number of such experiments has been performed. Hughes and Black (1973) have investigated the preference of hens for different types of battery cage floors. Preferences for litter or wire floors were reported by Hughes (1976b). Their spatial preferences and choice of group size were also investigated by Hughes (1975, 1977), while Dawkins (*loc. cit.*) has studied their choice between battery cage and an open run in a garden. Using operant techniques Baldwin and his co-workers have done a number of experiments with pigs allowing them to regulate the amount of light they receive, and to choose their own environmental temperature (Baldwin, 1974; Baldwin and Meese, 1977). There are a number of criticisms of this approach but, before proceeding to some of the general criticisms, a description of one of these experiments will show some of the pitfalls. Hughes (1976b) tested birds of the two laying strains mentioned in Chapter 9 (one a light-weight strain and the other a medium-weight hybrid) to determine whether they had any preferences between wire floors or solid floors covered with litter. Thirty six birds of each strain were used. Half of each strain had been reared on wire from hatching, and half on litter. Observations were made on how birds distributed their time between the two types of floor, each bird being observed over a 10-day period. The strain of the bird and the time of day of the observations were unimportant but the early rearing of the bird appeared to affect its choice, although not in all cases. In a second experiment birds with experience of either wire or litter were given a similar choice, but the method of giving the choice was different. Each bird was put into a runway, off which were two cages, one with a wire floor and one with litter. Then, having made its choice, the bird was obliged to spend 8 or 16 hours in that cage, with the result that none of the birds showed a preference for wire, although two showed no preference either way. Thus it can be seen that experimental methodology as well as early experience can affect the results. Furthermore, an animal will have to live in a social group and that too, could affect its choice.

In more general terms this type of experiment has been criticized by Duncan (1978) on the grounds that the tests cannot tell us about the absolute values of the choices, and that lower animals cannot be expected to weigh up the long-term consequences of their decisions. Thirdly, that the interpretation of the results is ambiguous. For example, the choice of one object in 10% of the tests may mean that that object is really unpopular or it may mean that for 10% of the time it is really important to the animal. In addition to these points raised by Duncan is the possibility that animals might make a choice which is harmful in the long term, rather like young people electing to smoke, despite widely publicized health risks. In view of this it might be reasonable to examine the factors that lead them to this choice. It may be that similar approaches in relation to animal choices will tell us more about the needs of the animals than the actual choice itself. Such an approach will require whole series of tests with controlled variables. Thus we might see that certain variables not only always influence the animal's choice, but influence it in a certain direction.

A second point made by Duncan (1978) that animals cannot be expected to make a rational choice needs investigation. The examples he cites relate only to poultry and, in our present state of knowledge, cannot unequivocally be applied to all farm livestock. The anticipation of its needs by a pig might be open to investigation: for example would a satiated pig choose a less popular pen with food and water over a popular pen without them if it had been trained that, having made its choice it had to stay there for 24 hours? If it did show anticipation of future thirst or hunger, could it anticipate other needs such as warmth or sleep?

Behavioural and physiological indices

In Chapter 9 we discussed how animals may respond to frustration. It will be recalled that detailed studies on stereotyped behaviour, which were discussed there, show that they occur when something is lacking in the environment, as in tethered sows (Fraser, 1975) or hens about to lay in a battery cage (Wood-Gush, 1972) or canaries in bare cages (Keiper, 1970), or when animals are stimulated by the presence of an inaccessible goal (Duncan and Wood-Gush, 1972b). It would seem therefore that the performance of these behaviour patterns is a symptom of a poor environment. However, it has been argued (Perry, 1978) that, while behaviour is the result of input of stimuli from the environment, all behaviour seen is part of the process of adaptation to the environment. Taken at one level, it may be that a stereotypy reduces or removes the frustration from the animal. However, it does not detract from the findings that some non-fixated stereotypies have been reduced by enrichment of the environment and that fixated stereotypies have been produced in fowls (Duncan and Wood-Gush, 1972b) and rats (Maier, 1949) by severe frustration. Nor does it diminish the moral argument against the environments that produce them. The evidence would support the view that stereotypies may be used as indices of unsatisfactory environments. With regard to other behavioural indices the picture is less clear.

In Chapter 9 we also discussed other behaviour patterns that occurred in conflict such as displacement activities, redirected activities and others. However, it would be very difficult to use these and other types of displacement behaviour as indices of welfare in our present state of knowledge. As some thwarting is inevitable in any environment, some displacement behaviour is to be expected, so there is the question as to how much is permissible? In any case a certain amount of stress is probably beneficial. Different stocks may show different types of behaviour on being thwarted. For example, of the two strains of hens observed by Wood-Gush (1972) during their pre-laying behaviour in battery cages, one performed vacuum nest building and the other stereotyped escape behaviour. The absence of certain behaviour patterns was suggested by Thorpe (1965) as being indicative of poor welfare, but as Dawkins (1976) has argued, one would not expect to see anti-predator behaviour in an intensive system and, furthermore, animals may substitute one type of behaviour for another. Apathy, by which is meant a lack of responsiveness to stimuli that normally would elicit a response, might be taken as an index, but it has yet to be clearly demonstrated in any species of farm livestock.

When animals are moved from one type of environment to another there is often a rapid and intense change of behaviour as when cattle are let out after a long period. Such behaviour could be a response to the sudden change of stimuli involving many stimuli rapidly impinging on it, or it could be due to a release of pent-up behaviour in response to relevant stimuli acting upon the animal after a period of deprivation, which has led to a need. If the second interpretation is correct, then this type of behaviour could be used to assess short-comings in a system, but obviously much research is needed into this point. The occurrence of deleterious behaviour such as navel sucking, tail biting or feather pecking is generally taken as being indicative of a faulty environment in the widest sense, including nutrition, but the remedial action taken is usually far removed from the ethological requirements of the animals. Feather pecking has been studied by a number of workers; its causation and development are complicated and often unpredictable (Hughes and Duncan, 1972; Cuthbertson, 1979). Analogous behaviour such as tail biting in pigs is unlikely to have simpler development and causation (Ewbank, 1969; Ekesbo, 1978).

Thorpe (1965) suggested that we should use distress behaviour as an index. However in most cases this is likely to be in response only to transitory treatment. Furthermore in the case of the chicken the issue is not clear cut, for Gentle and Wood-Gush (1976) found that decorticate fowls would squawk and struggle on being handled: 'distress' behaviour in this case does not necessarily involve the higher centres of the brain, but as it was less intense than in the normal fowl (Wood-Gush, unpublished), we may assume some cognitive involvement in the normal squawking of fowls. The use of telemetry to monitor physiological parameters in freely-moving animals, as is being done currently in fowls (Duncan, 1979), should help to correlate behaviour and the internal state, thus allowing better interpretation of the behaviour. In this technique a small transmitter is attached to the animal, or as in the case of telemetering heart-rate, implanted within the animal. This then transmits the electrocardiogram which is picked up on a receiver, the ECG signal analysed, and the heart rate then extracted. Similarly we need to know more about the relationships between overt behaviour and the endocrine symptoms of stress, particularly in relation to chronic stressors (see Chapter 11, Barnett *et al.*, 1981; Friend *et al.*, 1977).

The environmental features in relation to the animal's behaviour

A final method of assessing any husbandry system is to consider the major behavioural tendencies or 'drives' of the animal in relation to that system. Unfortunately until very recently we had very little ethological knowledge about our agricultural species, even in the case of the chicken, which has a long history of use as a laboratory animal for behavioural studies. The releasers or key stimuli that guide its behaviour and the underlying motivational mechanisms are generally also poorly understood. Wood-Gush (1973) attempted such an analysis of the battery cage as an environment for the expression of the major 'drives' of the laying hen. Considering first the reproductive behaviour of the hen, he indicated that it is unlikely that the sexual behaviour of the pullet will become overt until about the time of the first egg. At this time she will crouch to a variety of objects but in the absence of copulation the response wanes and there will very probably be no frustration. Since the development of sexual behaviour in the domestic hen has not been studied to the extent that it has in the cock, one has to extrapolate. In cocks that have been reared in visual isolation from other fowls, some will respond sexually to the hen on first meeting, whereas others behave aggressively at first and copulate only after some experience (Wood-Gush, 1958b). From other observations it appears that inexperienced cockerels will generalize to a wide variety of stimuli, if their own sexual vigour is high; if not they will, at first, behave aggressively towards hens and only after being presented with a large number of sexual crouches will they copulate. Jungle fowl cocks that have been deprived of sexual experience from hatching until about one year of age do not copulate readily (Kruijt, 1964), and much the same appears to be true of Brown Leghorn males (Wood-Gush, 1960). One may therefore assume that many pullets will possess only weak sexual motivation, if they lack sexual experience, and that this motivation will wane in the absence of potent releasers.

Nesting behaviour in the Jungle fowl and in a population of feral domestic fowl appears to be very similar. The hen starts by giving a particular call (Wood-Gush and Gilbert, 1969a) and this, in the Jungle fowl (Kruijt, 1964) and feral fowl (McBride, Parer and Foenander, 1969), entices the cock to her and he then leads her away to potential nest sites, one of which she eventually chooses as described in Chapter 6. In the domestic fowls that have been studied the hens give this call (Wood-Gush and Gilbert, 1969), but the cock takes no notice of it in some breeds and the hen proceeds to select her own nest site. The exact characteristics that govern whether a potential nest site will be chosen by a Jungle fowl hen have not yet been analysed. Judging by the description of sites used by Jungle fowl, the only common feature is relative isolation from flock mates (Murphy, 1969). Under semi-intensive conditions involving the use of trapnests, the pullet usually prefers to lay in a nest rather than on the floor, but there are some features of such nests that apparently disturb an inexperienced pullet; e.g. the trip wires on a trap-nest frighten many. However, the nests provided by modern agricultural conditions have enough stimuli to release nesting behaviour in hens about to lay. These stimuli have not been widely investigated, but an experiment with a small number of naive pullets suggests that relative darkness which has traditionally been considered to be important may be spurned by the inexperienced pullet (Wood-Gush and Murphy, 1970).

In the battery cage the hen normally goes through her nest- calling but the behaviour that precedes oviposition seems to depend on the stock. Wood-Gush (1975b) continued the study of the pre-laying behaviour of the two strains of fowls mentioned in Chapter 9. It will be recalled that in the battery cages the one, a White Leghorn strain, showed a great deal of stereotyped escape behaviour in the period before laying while the other, a heavy hybrid strain, sat quietly but performed vacuum nest-building behaviour. The main short-coming of the battery cage for the first strain appeared to be the wire floor, for if it was removed and a tray full of litter set in its place then the hens sat quietly. Since stereotypies are indicative of frustration there can be little doubt that the battery cage is a poor environment for the laying hen to lay in. Furthermore, we know nothing about the pre-laying behaviour of hens, kept in groups in cages, which may be even more disturbed. The other components of reproductive behaviour, viz. incubation and brooding behaviour, are uncommon in modern stocks and so need not be considered here.

In feeding behaviour some birds under modern husbandry conditions may suffer through being thwarted by other birds when they wish to eat. However, it appears that birds are able to consume enough food for growth in remarkably little time (Duncan *et al.*, 1970); probably 15 minutes is adequate feeding time under social conditions in a 14-hour day. The frustration of such birds is likely to be short-term and therefore they may wish to preen during this frustration (Duncan, 1970). Specific hungers are unlikely to be a major problem, but when a specific hunger is induced in fowls there appears to be an increase in general activity; in other species as well, some specific hungers lead to increased activity (Hughes, 1970a). Sodium-deficient calves show excessive licking, call more and lose appetite (Sly and Bell, 1979). In the case of the social behaviour of the fowl there is no specific physiological mechanism and therefore no predictable change in the behaviour except possibly during maturation. Nevertheless, all our agricultural animals have complex social systems, and the question 'how do modern agricultural methods affect these systems?' is a very large one, which needs to be reduced into simpler questions. What evidence have we that the restriction of social behaviour is harmful to, say, the fowl? Fowls kept in visual isolation from other fowls from hatching become extremely aggressive, and Jungle fowl hens raised in visual isolation from other birds spend much time trying to fight their own tails (Kruijt, 1964); self-mutilation may thus occur. Also during development these birds show excessive escape behaviour in response to small environmental changes, and such responses are much more prolonged than those of socially reared birds (Kruijt, 1964). Similarly, casual observations on domestic cocks raised in social isolation indicate that they are much more upset by changes in the environment than other birds. In the modern brooder a chick will have every opportunity to develop its social behaviour. Also in the single battery cage the bird will have visual and auditory communication with other birds. Indeed there is some evidence that they influence one another and show allelomimetic behaviour in feeding (Hughes, 1970b). In cages containing several birds, some lead very harassed lives. Frustration may be implicated, for when birds are frustrated in the presence of others there is a general increase in aggression (Duncan and Wood-Gush, 1971). However, it is interesting to note that an investigation by Hughes and Wood-Gush (1977) revealed that there may be less pecking in a battery cage than in a pen, presumably because the dominant bird in a cage, by her proximity, inhibits the others from pecking. Nevertheless there are few cages in which at least one bird is not harrassed.

It is unlikely that hens in battery cages are not going to be thwarted to some extent and it is certain that these birds will want to indulge in displacement preening. At some stocking densities this may be difficult for them but adherence to the recommended densities will allow them this outlet. Dust-bathing is also sometimes seen in hens in battery cages as a vacuum activity and in that environment cannot fulfil its function of maintaining good plumage. Vestergaard (1978) considers that this is a potential source of severe frustration in birds in battery cages.

It is often suggested that some of the troubles found in the poultry industry, such as feather-picking and cannibalism, are due to 'boredom'. Can one apply such a concept to an animal as primitive as a fowl, particularly if its behaviour is largely governed by releasers? Neuringer (1970) reported that pigeons, having learnt to peck a key for food, will persist in this behaviour even when the same type of food is freely available. Further experiments showed that this behaviour was not simply the 'carrying on' of a previously learnt habit. Similar findings have been reported in the fowl (Duncan and Hughes, 1972) and the authors suggest that this behaviour is due to experience, once having learnt to work for food the hen will continue to do so. However this is not an entirely satisfactory answer and the possibility of boredom as a motivating factor cannot be ruled out entirely (Wood-Gush *et al.*, 1982). It will be recalled that stereotypies are often performed by animals in bare environments, and head-flicking as a stereotyped behaviour pattern has been reported in fowls in battery cages (Bareham, 1972). All in all, the battery cage as built at present does not provide an environment conducive to the fulfilment of the hen's behavioural repertoire, but that does not mean that satisfactory cages cannot be designed.

This type of analysis has been carried further recently in the case of pigs. Stolba and Wood-Gush (in preparation) and Hutton *et al.* (1980) studied the behaviour of pigs under semi-natural conditions in an enclosure that contained mature trees, bushes, a marsh and a stream. They observed the behaviour in detail recording not only the behaviour patterns themselves, but how the pigs used the environment in carrying out their behaviour.

Much of the pigs' activities occurred at the border of the woodland and scrub area. Overnight they occupied a communal nest. All such nests constructed had certain common features, such as walls to protect the animals against the prevailing wind and an opening in the walls that allowed the animals a good line of vision as they lay in the nest. Furthermore, the nests were always very distant from the site where the pigs were fed. On rising in the morning the pigs tended to walk at least 7 metres before urinating and/or defaecating. This defaecation was always on paths and generally where it ran between bushes, unlike the cul-de-sac provided for dunging in many systems. The pigs frequently marked trees and this behaviour was highly allelomimetic. The pigs, which were small in number, formed very interesting relationships with very close bonds between certain animals, and new animals introduced to the area took a long time to be assimilated. Towards the end of the day adult members of the group indulged in nest building, adding material to the communal nest or rearranging what was there.

Several hours before parturition a sow would choose a nest site, well away from the communal nest and carry nesting material, often walking a large distance in the process (on one occasion a distance of 6 km was covered). She might even roll large logs into position as part of the walls. After parturition she would defend the nest against all pigs for several days, but after that might tolerate the presence of another post-partum sow with which she had a strong bond and the two litters would then share the nest. However, cross-suckling as practised in the industry was never seen. The piglets weaned themselves at about 15 weeks of age and the sows mostly came into oestrus and conceived while lactating (the group included an adult boar and a sub-adult male). Generally the behaviour of these Large White pigs, born and reared in an intensive system, once they had the appropriate environment resembled that of the European wild boar.

Stolba (1982) then observed the behaviour of similar groups in small paddocks with and without environmental features, such as bushes and found that in the small paddocks with environmental features the behavioural repertoire and sequences were very similar to those in the semi-natural conditions. However, in the same sized paddock without these features much of the behaviour was changed. This suggested to him that reduced space is not impediment to maintaining behaviour, it is the environmental features which contain the key stimuli and releasers for many of the behaviour patterns that are important. He therefore designed a housing system containing as many of the key features which he could identify, placing them in the same sort of spatial relationship as outside. Briefly these included:

1. A sleeping area well away from the feeding area, as had been found in the semi-natural area.
2. Open-fronted pens to resemble the forest border habitat.
3. A dunging corridor to resemble the paths between bushes.
4. A rooting area, since about 51% of the pigs' time was spent in rooting.
5. An activity area that contained a marking post and bedding material to be collected by the animals.

The close social bonds of the animals were not forgotten in the design, for the basic unit of the system is a group of 4 sows well known to each other. Each has a pen and all four pens are connected by the dunging corridor. Each pen has a sleeping area which can be shut off and used as a farrowing pen with farrowing rails. Furthermore, from the sleeping area the pigs have a good sight line as in the outdoor nests, and next to it is the activity area and outermost, the rooting area. Piglets born in the pen are kept there until the point of sale so that no other housing is required. Under these conditions the females tend to show synchrony of oestrus. The boar is brought in at 20 days after parturition and stays until all the females have been mated. By the time the next litter are born most of the previous litters are ready for sale or have been sold.

Although this system is still in the prototype stage is does provide a very good example of how a sound study of the behaviour of the animal can aid us in designing better housing. Closer investigation of the key stimuli guiding the main behaviour patterns would be of value, for it might allow greater control and use of space. Ultimately knowledge of these stimuli should enable us to trick the animal into being able to perform those behavioural patterns of importance to it, and for us to guide it sucessfully away from behaviour which we do not want, without causing frustration. Clearly, a knowledge and appreciation of the evolution of our agricultural animals together with information about the behaviour of feral populations will aid us in assessing housing systems and welfare.

Self-awareness

So far we have not considered the possibility of mental suffering. This would involve a degree of self-awareness in animals that few biologists would expect in animals, save in the higher primates. However, as Dawkins (1980) has pointed out, subjectivity (self-awareness) will have an evolutionary history and, on that score alone, one might expect it to be present to some degree in animals. She also points to the fact that mammals and birds have the basic neural apparatus for feeling pain and experiencing emotion. However, since in humans the perception of pain is influenced very much by psychological factors, so its perception by animals is difficult to assess. Nevertheless we must give the animal the benefit of any doubt in this matter. We saw in Chapter 12 that Rhesus monkeys, like humans, have different physiological responses to the anticipation of known and unknown unpleasant treatments. This suggests that their mental states could share some common features with man in this respect and, once again, raises the question of self-awareness in animals. One of the crucial points in any consideration of welfare, is the degree to which members of domesticated species can anticipate their future. Are they capable of having mental models which they can use to predict the behaviour of other members of their society? Can they judge their actions in relation to their environment, physical or social? On the other hand is their behaviour merely motivated by a few major drives? These are important questions to which we have no answers and until we do, the welfare of our farm livestock must be judged more pragmatically by the criteria discussed earlier.

POINTS FOR DISCUSSION

1. Assess the methods of intensive production of different classes of farm livestock (e.g. the breeding sow, the fattening pig, the dairy cow, veal calf, etc.) in relation to the welfare of the animals.
2. Discuss the welfare problems of extensively kept stock, e.g. the hill ewe.
3. Assess the welfare of an average suburban dog.
4. What are the main welfare problems involved with animals at abbatoirs?

FURTHER READING

Animal Welfare

Dawkins, M.S. (1980), *Animal Suffering: The science of animal welfare.* Chapman and Hall, London.

Duncan, I.J.H. (1974), A scientific assessment of welfare, *Proc. Br. Soc. Anim. Prod.*, 3, 9–19.

Dunn, P.G.C. (1980), Intensive livestock production: 'Costs exceed benefits,' *Vet. Rec.*, 106, 131–132.

Fölsch, D.W. (1978), *The ethology and ethics of farm animal production.* European Association for Animal Production, Basle, Switzerland. EAAP Publication no. 24.

Grandin, T. (1980), Designs and specifications for livestock handling equipment in slaughter plants, *Int. J. Stud. Anim. Prob.*, 1, 178–200.

McBride, G. (1968), Social organisation and stress in animal management, *Proc. Ecol. Soc. Aust.*, 3, 133–138.

Nehring, A. (1981), One answer to the confinement pig problem, *Int. J. Stud. Anim. Prob.*, 2, 256–259.

Various (1981) *Alternatives to intensive husbandry system*. UFAW, London.

Self-awareness in animals

Griffen, D.R. (1976), *The question of animal awareness. Evolutionary continuity of mental experience.* The Rockefeller University Press, New York.

Wood-Gush, D.G.M., Dawkins, M. and Ewbank, R. (eds) (1982), *Workshop on Self-Awareness in Domestic Animals.* UFAW, London.

15
ETHOLOGY AND PEST CONTROL

INSECT PESTS

In the control of insect pests, ethological knowledge can be, and is, used in several ways, e.g. the use of pheromones is practised. Some of these are true ethological releasers, acting on the nervous system to evoke a particular type of behaviour, involving attraction of the sexes, aggregation, or alarm behaviour. Another class of pheromones, called primer pheromones, act physiologically to alter the endocrine and reproductive systems of the receptor animal. Such pheromones can act either by physiological inhibition or enhancement (Matthews and Matthews, 1978). However, it is the former class that interests the ethologist most. Matthews and Matthews (*loc. cit.*) give a brief history of the early scientific investigation of insect pheromones. The role of chemical substances as sexual stimulants was first suggested by a German zoologist, von Siebold, about 140 years ago, but the idea was only pursued and developed fairly recently. In 1900 another zoologist, Mayer, investigated the mechanisms by which male promethea moths (*Hyalophora promethea*) are attracted to females. He noticed that female moths in a jar covered with mosquito netting attracted males from over 100 feet away. However, when the jar with the females was inverted and had sand packed around its mouth to prevent the transmission of scents, the males showed no interest. His next experiment indicated that visual contact was irrelevant; he loosely wrapped females in cotton wool, making them invisible, but males were still attracted and, indeed, tried to copulate with the cotton wool. His next experiment verified the involvement of chemical cues. Females were placed in a small wooden box in which air was blown in from one side and out of the other through a small chimney. Males were attracted not to the box, but to the chimney. Further experiments showed that the males perceived the chemical cues by means of their antennae.

Sex pheromones may be produced by either sex, although it is commonly the female that produces them. In some species, particularly among the beetles, both sexes may be attracted by the same scent. Today these pheromones have been used in apple orchards for surveying and monitoring the red-banded leaf roller, *Cargyrotaenia velutinana* (Roelofs, 1975) and citrus orchards against the California red scale, *Aonidiella aurantii* (Riehl *et al.*, 1980). They have also been used to reduce the numbers of cabbage loopers and pink bollworm moths in cabbage and cotton fields respectively (Matthews and Matthews, *loc. cit.*).

Other types of pheromones which are of potential use are those that lead to aggregation. An example is found in North America in the case of the Scolytid bark beetles which attack trees. The first pioneer beetles settle on an injured or subnormal tree, in response to volatile aldehydes or esters resulting from the abnormal enzyme activity of the injured tree. Thereupon, these pioneers discharge a pheromone which attracts both sexes. While this case of aggregation is connected with reproduction, other cases of aggregation involve immature animals, or animals of mixed ages (Matthews and Matthews, *loc. cit.*). The use of pheromones to control insect attack has many features to commend it. It is more specific than insecticides which tend to kill beneficial insects as well as harmful ones. Furthermore, since they are natural substances that regulate behaviour essential for survival of the species, the insects are less likely to become resistant to them than to conventional insecticides – although insects are remarkably adaptable. However, the sex pheromones tend to be effective in significantly reducing infestation only when infestation is light; when it is dense the males can probably find the females by sight (Marx, 1973).

Specific chemical stimuli are involved in the recognition of food and its acceptance by many insects. The potato beetle, *Leptinotarsa decemlineata*, is a good example in this respect. After hatching the larvae begin searching, guided mainly by vision, until they encounter a potato plant. They bite randomly on the substrate over which they are walking, and are attracted by short range olfactory stimuli emanating from the plant. If the relevant stimuli are encountered, the larva pierces the epidermis and if the taste is suitable, continues to feed. However, the releasing action of some pheromones is so powerful that some species can be guided to eat inappropriate plants if these have been sprayed with the appropriate chemical. For example, sinigrin, a mustard oil glucoside characteristic of the plant family *Cruciferae*, acts as a releaser for feeding in the turnip and cabbage aphids. When the leaves of ten non-host plants were treated with sinigrin, the aphids readily ate the plants (Nault and Styer, 1972). On the other hand, some plants are protected by specific chemicals. Table 15.1 shows how various alkaloids deter the larvae of the Colorado potato beetle, *Leptinotorsa decemineata*, from feeding on various members of the *Solanaceae* family, while the alkaloid of the potato has no such effect (cited by Matthews and Matthews, *loc. cit.*).

However, it should not be thought that only herbivorous insects or, indeed, only insects find their food by means of specific chemicals. One of the earliest examples cited in the ethological literature is a species of tick that in the imaginal stage feeds on mammals. Guided by

Table 15.1 The effects of various alkaloids from Solanaceae as deterrents to feeding by the Colorado potato beetle, *Leptinotarsa decemlineata*. (From Matthews and Matthews, 1978.)

Chemical	Plant origin	Degree of inhibition*
Solanine	*Solanum tuberosum*	−
Chaconine	*S. tuberosum, S. chacoense*	−
Demissine	*S. demissum, S. jamesii*	+++
Leptine	*S. chacoense*	+++
Soladucine	*S. dulcamara*	I I
Solacauline	*S. acuale, S. caulescens*	++
Solamargine	*S. aviculare, S. sodomeum*	−
Solanigrine	*S. nigrum*	++
Solasonine	*S. sodomeum, S. carolinense*	−
Tomatine	*Lycopersicon esculentum*	+++
Capsaicin	*S. capsicum*	++
Nicotine	*Nicotiana tabacum, N. rustica*	Toxic
Nicandrenone	*Nicandra physalodes*	+++
Atropine	*Atropa belladonna*	+
Scopolamine	*Datura* spp.	++

* +++, Strong; ++, moderate; +, slight; −, no deterrent effects.

light cues it climbs up a grass stem or a bush. Once there it rests until an appropriate host brushes it off, or passes below, in which case the tick drops onto it. The releaser for the tick to drop is butyric acid which emanates from the host (Von Uëxküell, 1957). Aggregation pheromones have also been reported in ticks (Petney and Bull, 1981; Trevorrow *et al.*, 1977).

The use of biological control to combat insect pests, that is to employ their natural predators or parasites, is advocated by some. Very often where they have been applied they have been highly successful. Huffaker and Smith (1980) state that out of 327 attempts to control insect pests by biological control, 102 were completely successful and 144 cases showed substantial success. The process requires as much knowledge as possible about the behaviour and ecology of the species involved. Apart from knowledge about breeding and feeding behaviour of pests and their enemies, ethological studies can contribute to understanding the factors that lead to swarming and movement in general. Understandably these are important factors in the development of management systems against pests (Armbrust *et al.*, 1980). Another method of insect pest control that has been used is autocidal control. This method, in which the pest population is saturated with sterile males, has been used very successfully in the USA against the screw worm (the larva of *Cochliomyia hominivorax*, a blowfly) which attacks cattle and sheep. However it is necessary to know the mating behaviour of the species and how the treated males fare in different populations which may be very diverse (Richardson *et al.*, 1982).

The development of resistance to insecticides may sometimes involve behavioural adaptations. Broom (1981) cites a number of cases. Some houseflies and some malaria-carrying mosquitoes are resistant to DDT because they will not settle on DDT treated surfaces for long enough to absorb a lethal dose. Also the susceptibility of an insect species, or strain of a species, may depend on the general level of activity of that species or strain. Conversely, one strain of grain weevil, which was resistant to the insecticide pyrethrum, was found to be less active, suggesting that its immunity was due to the fact that it encountered less of the insecticide per unit of time. The beetle, *Tribolium castaneum*, which infests grain stores, has been found to actively avoid surfaces treated with some insecticides and to crawl into crevices where the insecticide has not penetrated.

ETHOLOGY AND OTHER PESTS

Not all pests are insects. In fact only about 1% of the million or so species of insects are pests; indeed some are beneficial to agriculture. Birds in some areas and under some conditions can be formidable pests. In parts of Africa, the African weaver bird, *Quelea quelea*, can decimate crops of millet or sorghum. However, its main food items are grass seeds and the domestic crops are only used at times of the year when its natural food sources decline – a feature which is found in many bird pests. Killing a proportion of the pest seems generally to have very little effect on numbers as a whole, for it merely seems to kill off the proportion that would die anyway. In South Africa a million *Quelea* were killed by aerial spraying in a particular area, but even a reduction of that order had very short-term effects. Generally, ecological or managerial measures seem to be more effective. For example, a variety of the crop can sometimes be used that grows when the pest's natural food sources are abundant, or, as in the case of bullfinch predation on pear trees, a shorter variety that can be covered by nets is now being used (Flegg, 1980).

Ethology may help to design better devices for scaring birds. Vocal scaring by playing the distress calls of the pest species has been used in a variety of situations from central urban sites to airfields. A very successful case is described by Spanier (1980). In Israel, night herons, (*Nycticorax nycticorax*), inflict heavy losses at fish farms, but the playing of their alarm calls repelled more than 80% of the would-be marauders. Furthermore, the population was almost even after weeks of nightly broadcasts. Generally however, this type of scarer has had limited success; after the initial effect the birds have habituated to the relayed calls. Brémond (1980) has suggested a number of interesting points in relation to this type of scarer. He points out that it is known that some birds vary their territorial song in relation to their degree of aggressiveness. In all cases a group of parameters of the vocal message remain without modification to ensure the transmission of information concerning species identification and motivational state, whilst another group is altered to express the level of motivational state. He suggests that when distress calls are played and have only limited effect, it may be that they are in fact a mixture of distress and alarm calls, the proportion of which is wrong. He proposes the development of vocal super-stimuli, and cites one case which has already been discovered. The common American crow (*Corvus brachyrhynchos*) has an 'assembly' call consisting of cawing in a particular structured form. There is also an unstructured form of cawing used for mobbing and dispersion. By studying the different types of cawing, workers found that a high and increasing rate of emission of caws promotes assembly and, from the two types of cawing, have manufactured an assembly call which is more efficient in causing assembly than the natural one. Although Brémond goes on to discuss the possibility of manufacturing acoustic super-stimuli that would act as rejection markers in order to keep birds away from a crop, an assembly call that lured them to a trap or, say, an alternative roosting site might also be considered in certain cases. Slater (1980) suggests that to prevent, or to delay, habituation the stimulus should be played infrequently, have variation, and be accompanied by reinforcement such as shooting or presentation of a bird of prey. Any shooting should, of course, be done in close juxta-position to the vocal stimulus so that the birds connect the two, or if a bird of prey is used it should be one which they are likely to have encountered in the wild.

Visual bird scarers have a long history and Inglis (1980) has analysed the effectiveness of modern ones on the market. He finds that most do not use features that naturally frighten the bird, but rely largely on novelty and mainly employ features that are not normally encountered by the birds, such as rotating orange vanes. Such novel scarers, however, with repeated exposure should become attractive to the birds, particularly as they are always coupled with positive reinforcement in the form of the crop! Any scarers, visual or otherwise should at some time be coupled with aversive stimuli so that they become warning stimuli. From experiments with human-like scarers, Inglis has found that models with flapping arms are the most effective, possibly because the flapping arms resemble the flapping wings of a bird of prey. He advocates that if models of raptorial birds are used, then there will be more success if the model is visible for short periods only, is mobile at those times and is shown carrying a model of the pest species. Furthermore its appearance should be coupled with the alarm calls of the pest species. In cases in which one is dealing with a single bird species, the presentation of a model of the pest in an alarm posture or unusual posture may be effective, although this has not always been the case.

The brown rat is, of course, one of the most common pests of farm buildings. Behavioural studies have shown that they display a high degree of neophobia in their feeding behaviour (Barnett, 1975). Any new food item is initially shunned. Only after many encounters with a new source of food will they begin to eat from it and when they do, they tend to take very small amounts. This means that any bait should be presented many times, and only when there are obvious signs of large amounts of it having been eaten, should any poison be added. If the poison is added too soon, when the rats are taking only a small amount, they might only eat enough to make them ill, and then form a conditioned avoidance response to that bait. However, conditioned avoidance has been used in an attempt to control certain pests.

In an interesting study Gustavson *et al.* (1976) used coyotes which sometimes kill lambs. Coyotes were presented under laboratory conditions with a choice between rabbits or chickens as prey. After it was established that rabbits were the preferred prey in all cases, each animal was presented with rabbit flesh tainted with lithium chloride, an emetic in small doses. After this treatment (see Table 15.2) attacks on rabbits in the choice situation declined very markedly. A field trial was followed in which sheep meat on a farm was similarly treated and following this it appeared that the lamb deaths due to killing by coyotes dropped by 30–60%, compared to the previous three years. However, as yet there is not enough evidence to say whether this technique is thoroughly successful or not (Burns, 1980).

Table 15.2 The effect on the predatory behaviour of coyotes from treating the flesh of a preferred source of prey (rabbit) with lithium chloride (Adapted from Gustavson *et. al.*, 1976.)

Coyote	Tests before treatments		Prey flesh treated with LiCl	Tests after treatments	
	Rabbit	*Chicken*		*Rabbit*	*Chicken*
1	Attacked	Attacked	Rabbit	–	Attacked
2	Attacked	Attacked	Rabbit	–	Attacked
3	Attacked	Attacked	Rabbit	–	Attacked
4	Attacked	Attacked	Rabbit	–	Attacked
5	Attacked	–	Rabbit	–	Attacked

CONCLUSIONS

As in many areas of biology the effective biological control of pests depends on both ecology and ethology. While ecology is the major contributor in this sphere, ethological techniques and knowledge can supply greater insight into a number of areas. Studies on social, reproductive, and feeding behaviour of the pest are obviously relevant. Analysis of the stimuli that govern individual and group movement are also important. Ethological studies can further help to elucidate certain cases of resistance which, as we saw previously, can be behavioural rather than physiological. In addition, as stated, the analysis of the interactions of the pest and its natural enemies also requires ethological study. With such knowledge, effective management systems for all pests, from invertebrates to mammals, should be possible without the widespread use of pesticides or other dangerous practices that challenge the ecological balance.

POINTS FOR DISCUSSION

1. State why a knowledge of the social behaviour of a pest species may aid control of that pest.
2. Do you consider that there are any differences in principle in the control of, say, aphids and rabbits?
3. Discuss the pros and cons of using pheromones against insect pests.

FURTHER READING

Green, R.E. (1980), Food selection by skylarks and grazing damage to sugar beet seedlings, *J. Appl. Ecol.*, **17**, 613–630.
Henderson, B.A. (1979), Regulation of the size of the breeding population of the European rabbit, *Oryctolagus cuniculus*, by social behaviour, *J. Appl. Ecol.*, **16**, 383–392.
Huffaker, C.B. (ed.) (1980), *New Technology of Pest Control*. John Wiley & Sons, Chichester.
Loudon, A.S.I. (1978), The control of Roe deer populations: A problem in forest management, *Forestry*, **51**, 73–83.
Mitchell, E.R. (1981), Management of insect pests with semiochemicals, Plenum Press, London.
Wright, E.N., Inglis, I.R. and Feare, C.J. (eds.) (1980), *Bird problems in agriculture. Proceedings of a Conference 'Understanding bird problems'.* BCPC Publication, Croydon.

REFERENCES

Alexander, G. and Shillito, E.E. (1977), The importance of odour, appearance and voice in maternal recognition of the young in merino sheep (*Ovis aries*), *Appl. Anim. Ethol.*, **3**, 127–135.

Alexander, G. and Williams, D. (1964), Maternal facilitation of suckling drive in new born lambs, *Science*, **146**, 665–666.

Alexander, G. and Williams, D. (1966), Teat-seeking activity in lambs during the first hours of life, *Anim. Behav.*, **14**, 166–178.

Allee, W.C., Collias, N.E. and Lutherman, C. (1939), Modification of the social order in flocks of hens by the injection of testosterone propionate, *Physiol. Zool.*, **12**, 412–420.

Allison, T. and Cicchetti, D.V. (1976), Sleep in Mammals: Ecological and constitutional correlates, *Science*, **194**, 732–735.

Amlaner, C.J.Jr. and McFarland, D.J. (1981), Sleep in the Herring gull (*Larus argentatus*), *Anim. Behav.*, **29**, 551–556.

Andersson, B. (1971), Thirst and brain control of water balance, *Am. Scientist*, **59**, 408–415.

Andersson, B and Wyrwicka, W. (1957), Electric stimulation of the 'Drinking Area', *Acta Physiol. Scand.*, **41**, 194–198.

Andrew, R.J. (1956), Some remarks on behaviour in conflict situations with special reference to Emberiza spp., *Br. J. Anim. Behav.*, **4**, 41–45.

Andrew, R.J. (1966), Precocious adult behaviour in the young chick, *Anim. Behav.*, **14**, 485–500.

Archer, J. (1976), The organization of aggression and fear in vertebrates, in *Perspectives in Ethology*, (Bateson, P.P.G. and Klopfer, P.H., eds.), **2**, 231–298.

Archer, J. (1979), *Animals Under Stress*. Studies in Biology No. 108. Edward Arnold, London.

Armbrust, E.J., Pass, B.C., Davis, D.W., Helgesen, R.G., Manglitz, G.R., Pienkowski, R.L. and Summers, C.G. (1980), General accomplishments toward better insect control in alfalfa, in *New Technology of Pest Control* (Huffaker, C.B., ed.), John Wiley & Sons, Chichester, pp. 187–216.

Arnold, G.W. (1970), Regulation of food intake in grazing animals, in *Physiology of Digestion and Metabolism in the Ruminant*. Proc. 3rd Internat. Symp. Cambridge, England (Phillipson, A.T., ed.). Oriel Press, Newcastle-upon-Tyne, pp. 264–276.

Arnold, G.W. (1977), An analysis of spatial leadership in a small field in a small flock of sheep, *Appl. Anim. Ethol.*, **3**, 263–270.

Arnold, G.W., Boundy, C.A.P., Morgan, P.D. and Bartle, G. (1975), The roles of sight and hearing in the lamb in the location and discrimination between ewes, *Appl. Anim. Ethol.*, **1**, 167–176.

Arnold, G.W. and Dudzinski, M.L. (1978), *Ethology of Free-ranging Domestic Animals*. Elsevier Scientific Publishing Co., Amsterdam.

Baille, C.A. and Forbes, J.M. (1974), Control of feed intake and regulation of energy balance in ruminants, *Physiol. Rev.*, **54**, 160–214.

Baker, N. (1981), *The characterization of chronic stereotyped behaviour in stalled sows (Sus scrofa): with special reference to function, causation and effects on normal behaviour*. Unpublished Honours thesis, Dept. of Zoology, University of Edinburgh.

Balch, C.C. (1955), Sleep in ruminants, *Nature, Lond.*, **175**, 940–941.

Baldwin, B.A. (1974), Behaviour thermoregulation, in *Heat Loss from Animals and Man* (Monteith, J.L. and Mount, L.E., eds), Butterworths, London, pp. 97–117.

Baldwin, B.A. and Meese, G.B. (1977), Sensory reinforcement and illumination preference in the domesticated pig, *Anim. Behav.*, **25**, 497–507.

Baldwin, B.A. and Stephens, D.B. (1973), The effects of conditioned behaviour and environmental factors on plasma corticosteroid levels in pigs, *Physiol. Behav.*, **10**, 267–274.

Banks, E.M. (1965), Some aspects of sexual behaviour in domestic sheep (*Ovis aries*), *Behaviour*, **23**, 249–279.

Banks, E.M. (1977), Vertebrate social organization, in *Internat. Encyclopaedia of Psychiatry, Psychology, Psychoanalysis and Neurology*, Aesculapius Publishers Inc., pp. 360–364.

Bareham, J.R. (1972), Effects of caged and semi- intensive deep litter pens on the behaviour, adrenal response and production in two strains of laying hens, *Br. Vet. J.*, **128**, 153–162.

Bareham, J.R. (1975), The effect of lack of vision on suckling behaviour of lambs, *Appl. Anim. Ethol.*, **1**, 245–250.

Barnett, J.L., Cronin, G.M. and Winfield, C.G. (1981), The effects of individual and group penning of pigs on total and free plasma corticosteroids and the maximum corticosteroid binding capacity, *Gen. Comp. Endocrin.*, **44**, 219–225.

Barnett, S.A. (1958), Physiological effects of 'social stress' in wild rats. 1: The adrenal cortex, *J. Psychosomat. Res.*, **3**, 1–11.

Barnett, S.A. (1975), *The Rat: A Study in Behaviour*. Univ. of Chicago Press, London.

Bastock, M. (1967), *Courtship, a Zoological Study*. Heinemann Educational Books Ltd., London.

Bateson, P.P.G. (1977), The development of play in cats, *Appl. Anim. Ethol.*, 4, 290.

Batty, J. (1978), Plasma levels of testosterone and male sexual behaviour in strains of house mouse (*Mus musculus*), *Anim. Behav.*, 26, 339–348.

Beach, F.A. (1976), Sexual attractivity, proceptivity and receptivity in female mammals, *Hormones and Behav.*, 7, 105–138.

Beach, F.A. and Jordon, L. (1956), Sexual exhaustion and recovery in the male rat, *Quart. J. Exp. Biol.*, 8, 101–133.

Bekoff, M. (1976), Animal play: problems and perspectives, in *Perspectives in Ethology*, 2 (Bateson, P.P.G. and Klopfer, P.H., eds). Plenum Press, London, pp. 165–188.

Bell, F.R. (1971), Hypothalamic control of food intake, *Proc. Nutr. Soc.*, 30, 103–109.

Berger, J. (1978), Group size, foraging and anti-predator ploys: An analysis of Bighorn sheep decisions, *Behav. Ecol., Sociobiol.*, 4, 91–99.

Berger, J. (1979), Social ontogeny and behavioural diversity: consequences for Bighorn sheep (*Ovis canadensis*) inhabiting desert and mountain environments, *J. Zool. Soc. Lond.*, 188, 251–266.

Berlyne, D.E. (1960), *Conflict, Arousal and Curiosity*. McGraw-Hill, London.

Best, O.H. and Taylor, N.B. (1955), *The Physiological Basis of Medical Practice*. Williams and Wilkins, Baltimore.

Beuving, G. and Vonder, G.M.A. (1978), Daily rhythm of corticosterone in laying hens and the influence of egg laying, *J. Reprod. Fert.*, 51, 169–173.

Bindra, D. (1959), The interpretation of the 'displacement' phenomenon, *Br. J. Psychol.*, 50, 263–268.

Blass, E.M. (1973), Cellular dehydration thirst: Physiological, neurological and behaviour correlates, in *The Neuropsychology of Thirst: New Findings and Advances in Concepts* (Epstein, A.N., Kissileff, H.R. and Stellar, E., eds). Winston VH & Sons, Washington DC, pp. 37–72.

Blundell, J.E. (1976), The C.N.S. and Feeding. Group Report, in *Appetite and Food Intake: Report of the Dahlem Workshop on Appetite and Food Intake* (Silverstone, T., ed.) Life Sciences Research Report 2, Berlin. pp. 109–128.

Booth, D.A. (1972), Postabsorptively induced suppression of appetite and the energostatic control of feeding, *Physiol. Behav.*, 9, 199–202.

Bouissou, M.F. (1972), Influence of body weight and presence of horns on social rank in domestic cattle, *Anim. Behav.*, 20, 474–477.

Bouissou, M.F. and Andrieu, S. (1978), Establissement des relations preferentielles chez les bovins domestiques, *Behaviour*, **64**, 148–157.

Bray, G.A. (1976), Peripheral metabolic factors in the regulation of feeding, in *Appetite and Food Intake: Report of the Dahlem Workshop on Appetite and Food Intake* (Silverstone, T., ed.) Life Sciences Research Report 2, Berlin, pp. 109–128.

Bremond, J.-C. (1980), Prospects for making acoustic super-stimuli, in *Bird Problems in Agriculture*. Proceedings of a Conference 'Understanding Bird Problems' (Wright, E.N., Inglis, I.R. and Feare, C.J., eds.). BCPC Publication, Croydon, pp. 115–120.

Brobeck, J.R. (1957), Neural basis of hunger, appetite and satiety, *Gastroenterology*, **32**, 169–174.

Broom, D.M. (1981), *Biology of Behaviour*. Cambridge University Press, Cambridge.

Broom, D.M. and Leaver, J.D. (1978), Effects of group rearing or partial isolation on later social behaviour of calves, *Anim. Behav.*, **26**, 1255–63.

Brownlee, A. (1954), Play in domestic cattle: an analysis of its nature, *Brit. Vet. J.*, **110**, 48–68.

Burns, R.J. (1980), Effect of lithium chloride in coyote pup diet, *Behav. Neurol. Biol.*, **30**, 350–356.

Candland, D.K. (1969), Discriminability of facial regions used by the domestic chicken in maintaining the social dominance order, *J. Comp. Physiol. Psychol.*, **69**, 281–285.

Cannon, W.B. (1934), Hunger and thirst, in *Handbook of General Experimental Psychology*. Clark University Press, Worcester, Massachusetts.

Cannon, W.B. and Washburn, A.C. (1912), An explanation of hunger, *Am. J. Physiol.*, **29**, 441–454.

Carbaugh, B.T., Schein, M.W. and Hale, E.B. (1962), Effects of morphological variations of chicken models on sexual responses of cocks, *Anim. Behav.*, **10**, 235–238.

Cheng, Mei-Fang (1975), Induction of incubation behaviour in male Ring Doves (*Streptopelia risoria*): A behavioural analysis, *J. Reprod. Fert.*, **42**, 267–276.

Christian, J.J. (1950), The adreno-pituitary system and population cycles in mammals, *J. Mammal.*, **31**, 247–259.

Christian, J.J. (1955), Effect of population size on the adrenal glands and reproductive organs of male mice in populations of fixed size, *Am. J. Physiol.*, **182**, 292–300.

Collias, N.E. (1943), Statistical analysis of factors which make for success in initial encounters between hens, *Am. Nat.*, **77**, 519–538.

Collias, N.E. (1952), The development of social behaviour in birds, *Auk*, **69**, 127–159.

Collias, N.E. and Joos, M. (1953), The spectrographic analysis of sound signals of the domestic fowl, *Behaviour*, 5, 175–188.

Craig, J.V. (1980), *Domestic Animal Behaviour*, Prentice-Hall, New York.

Craig, J.V., Casida, L.E. and Chapman, A.B. (1954), Male infertility associated with lack of libido in the rat, *Am. Nat.*, **88**, 365–372.

Craig, J.V., Ortman, L.L. and Guhl, A.M. (1965), Genetic selection for social dominance ability in chicks, *Anim. Behav.*, **13**, 114–131.

Crofton, H.D. (1958), Nematode parasite populations in sheep on lowland farms, *Parasitology*, **48**, 235–260.

Crook, J.H. and Butterfield, P.A. (1968), Effects of testosterone and luteinizing hormone on agonistic and nest building behaviour of *Quelea quelea, Anim. Behav.*, **16**, 370–384.

Cruze, W.W. (1935), Maturation and learning in chicks, *J. Comp. Psychol.*, **19**, 371–409.

Cuthbertson, G.J. (1979), *An ethological investigation into feather pecking*, Unpublished PhD thesis, University of Edinburgh.

Dantzer, R., Arnone, M. and Mormede, P. (1980), Effects of frustration on behaviour and plasma corticosteroid levels in pigs, *Physiol. Behav.*, **24**, 1–4.

Davies, C.M. (1928), Self-selection of diet by newly-weaned infants, *Am. J. Dis. Child*, **36**, 651–679.

Davis, D.E. (1957), Aggressive behaviour in castrated starlings, *Science*, **126**, 353.

Davis, D.E. and Domm, L.V. (1943), The influence of hormones on the sexual behaviour of the fowl, in *Essays in Biology*. University of California Press, pp. 171–181.

Dawkins, M. (1976), Towards an objective method of assessing welfare, *Appl. Anim. Ethol.*, **2**, 245–254.

Dawkins, M. (1977), Do hens suffer in battery cages? Environmental preferences and welfare, *Anim. Behav.*, **25**, 1034–1046.

Delius, J. (1967), Displacement activities and arousal, *Nature, Lond.*, **214**, 1259–1260.

Denenberg, V.H. (1973), Developmental factors in aggression, in *The Control of Aggression* (Knutson, J.F., ed.). Aldine Publishing Co., Chicago, pp. 41–58.

Denton, D. (1967), Salt appetites, in *Handbook of Physiology*, Sect. 6, vol. 1. Alimentary Canal (Code, C.F. and Heidel, W., eds) Amer. Physiol. Soc., Washington DC, pp. 443–459.

Desforges, M.F. (1973), *Behaviour studies on the Aylesbury domestic duck and some comparisons with wild Mallard (Anas platyrhynchos)*, Unpublished PhD thesis, University of Edinburgh.

Desforges, M.F. and Wood-Gush, D.G.M. (1975a), A behavioural comparison of domestic and mallard ducks. Spatial relations in small flocks, *Anim. Behav.*, **23**, 698–705.

Desforges, M.F. and Wood-Gush, D.G.M. (1975b), A behavioural comparison of domestic and mallard ducks. Habituation and flight reactions, *Anim. Behav.*, **23**, 692–697.

Desforges, M.F. and Wood-Gush, D.G.M. (1976), Behavioural comparison of Aylesbury and mallard duck: sexual behaviour, *Anim. Behav.*, **24**, 391–397.

Dickson, D.P., Barr, G.R. and Wieckert, D.A. (1967), Social relationships of dairy cows in a feed lot, *Behaviour*, **29**, 195–203.

Dilger, W.C. (1962), The behaviour of love birds, *Sci. Am.*, **206**, 88–98.

Dobie, M.J. (1979), *A study of some aspects of the social behaviour of the Blackface ewe under hill conditions*, Unpublished Honours thesis, The School of Agriculture, The University of Edinburgh.

Dollard, J., Doob, L.W., Miller, N.E., Mowrer, O.H. and Sears, R.R. (1939), *Frustration and Aggression*, Yale University Press, New Haven.

Drummond, J.C. and Wilbraham, A. (1957), *The Englishmans Food. Five Centuries of English Diet.* Jonathan Cape, London.

Duncan, I.J.H. (1970), Frustration in the fowl, in *Aspects of Poultry Behaviour* (Freeman, B. and Gordon, R.F., eds). British Egg Marketing Board Symp. No. 6, pp. 15–32.

Duncan, I.J.H. (1978), The interpretation of preference tests in animal behaviour, *Appl. Anim. Ethol.*, **4**, 197–200.

Duncan, I.J.H. (1979), Some studies on heart-rate and behaviour in the domestic fowl, *Appl. Anim. Ethol.*, **5**, 294–295.

Duncan, I.J.H. and Filshie, J.H. (1980), The use of radiotelemetry devices to measure temperatures and heart rate in the domestic fowl in *A Handbook on Biotelemetry and Radio Tracking* (Amlaner, C.J., McDonald, D.W., eds). Pergamon Press, London, pp. 579–588.

Duncan, I.J.H., Horne, A.R., Hughes, B.O. and Wood-Gush, D.G.M. (1970), The pattern of food intake in the female Brown Leghorn fowl as recorded in a Skinner Box, *Anim. Behav.*, **18**, 245–255.

Duncan, I.J.H. and Hughes, B.O. (1972), Free and operant feeding in the domestic fowls, *Anim. Behav.*, **20**, 775–777.

Duncan, I.J.H., Savory, C.J. and Wood-Gush, D.G.M. (1978), Observation on the reproductive behaviour of domestic fowl in the wild, *Appl. Anim. Ethol.*, **4**, 29–42.

Duncan, I.J.H. and Wood-Gush, D.G.M. (1971), Frustration and aggression in the domestic fowl, *Anim. Behav.*, **19**, 500–504.

Duncan, I.J.H. and Wood-Gush, D.G.M. (1972a), An analysis of displacement preening in the domestic fowl, *Anim. Behav.*, **20**, 68–71.

Duncan, I.J.H. and Wood-Gush, D.G.M. (1972b), Thwarting of feeding behaviour in the domestic fowl, *Anim. Behav.*, **20**, 444–451.

Duncan, I.J.H. and Wood-Gush, D.G.M. (1974), The effects of a rauwolfia tranquilizer on stereotyped movements in frustrated domestic fowl, *Appl. Anim. Ethol.*, **1**, 67–76.

Edney, T.N., Kilgour, R. and Bremner, K. (1978), Sexual behaviour and reproductive performance of ewe lambs at and after puberty, *J. Agric. Sci. Camb.*, **90**, 83–91.

Edwards, C.A. (1980), *Behavioural interactions between dairy cows and their newborn calves with relation to calf serum immunoglobulin levels.* Unpublished PhD thesis, University of Reading.

Eisener, E. (1960), The relationship of hormones to the reproductive behaviour of birds, referring especially to parental behaviour: a review, *Anim. Behav.*, **8**, 155–179.

Ekesbo, I. (1978), Ethics, ethology and animal health in modern Swedish livestock production, in *The Ethology and Ethics of Farm Animal Production* (Fölsch, D.W., ed.). EAAP Publication No. 24, pp. 46–50.

Ewbank, R. (1969), Behavioural implications of intensive animal husbandry, *Outlook in Agric.*, **6**, 41–46.

Ewbank, R. and Meese, G.B. (1974), Individual recognition and the dominance hierarchy in the domestic pig: the role of sight, *Anim. Behav.*, **22**, 473–481.

Fagan, R.M. (1976), Exercise, play and physical training in animals, in *Perspectives in Ethology*, 2. (Bateson, P.P.G. and Klopfer, P.H., eds) Plenum Press, London, pp. 165–188.

Falk, J.L. (1977), The origin and functions of adjunctive behaviour, *Anim. Learn. Behav.*, **5**, 325–335.

Feist, J.D. (1971), *Behaviour of feral horses in the Pryor Mountain Wild Horse Range*, Master's thesis, University of Michigan, Ann. Arbor.

Flegg, J.J.M. (1980), Biological factors affecting control strategy, in *Bird Problems in Agriculture.* Proceedings of a Conference 'Understanding Bird Problems' (Wright, E.N., Inglis, I.R. and Feare, C.J., eds). BCPC Publications Croydon, pp. 7–19.

Fletcher, T.J. and Short, R. (1974), Restoration of libido in castrated red deer (*Cervus elaphus*) with oestradiol-17β, *Nature, Lond.*, **248**, 616.

Fox, M.W. (1969), The anatomy of aggression and its ritualization in Canidae: a developmental and comparative study, *Behaviour*, **35**, 242–258.

Fox, M.W. (1971), *Behaviour of Wolves, Dogs and Related Canids.* Jonathan Cape, London.

Fox, M.W. and Bekoff, M. (1975), The behaviour of dogs, in *The Behaviour of Domestic Animals*, 3rd edn (Hafez, E.S.E., ed.). Bailliere Tindall, London, pp. 370–409.

Fradrich, H. (1974), A comparison of behaviour in the Suidae. Paper No. 6, in *The Proc. of the Conf. on the Behaviour of Ungulates and Its Relation to Management* (Geist, V. and Walther, F., eds). IUCN, Morges, Switzerland.

Francis-Smith, K. (1979), *Studies on the Feeding and Social Behaviour of Domestic Horses*, Unpublished PhD thesis, University of Edinburgh.

Fraser, D. (1975), The effect of straw on the behaviour of sows in tether stalls, *Anim. Prod.*, **21**, 59–68.

Freekes, F. (1971), *'Irrelevant' Ground-pecking in Agonistic Situations in Burmese Red Jungle Fowl (Gallus gallus spadiceus)*, Doctoral thesis, The University of Groningen, The Netherlands.

Friend, T.H., Polan, C.E., Gwazdouskas, F.C. and Heald, C.W. (1977), Adrenal glucocorticoid response to exogenous adrenocorticotrophin mediated by density and social disruption in lactating cows, *J. Dairy Sci.*, **60**, pp. 1958–1963.

Freyman, R. (1963), Follow-up study of enuresis treated with a bell apparatus, *J. Child Psychol. Psychiat.*, **4**, 199–206.

Geist, V. (1971), *Mountain Sheep. A study in Behaviour and Evolution.* University of Chicago Press, London.

Gentle, M.J. (1974), Changes in habituation of the EEG to water deprivation and crop loading in *Gallus domesticus*, *Physiol. Behav.*, **13**, 15–19.

Gentle, M.J. and Wood-Gush, D.G.M. (1976), Observations on the behaviour of *Gallus domesticus* following telencephalic removal, *Biol. Behav.*, **1**, 223–231.

Gibbs, J., Young, R.C. and Smith, G.P. (1973), Cholecystokinin decreases food intake in rats, *J. Comp. Physiol. Psychol.*, **84**, 488–495.

Glendinning, S.A. (1977), The behaviour of sucking foals, *Brit. Vet. J.*, **133**, 192.

Glickman, S.E. (1958), Effects of peripheral blindness on exploratory behaviour, *Can. J. Psychol.*, **12**, 45–51.

Goodall, J. (1964), Tool using and aimed throwing in a community of free-living chimpanzees, *Nature, Lond.*, **201**, 1264–1266.

Gordon, J.G. and Tribe, D.E. (1951), The self-selection of diet by pregnant ewes, *J. Agric. Sci.*, **41**, 187–190.

Goy, R.W. and Jakway, J.S. (1959), The inheritance of patterns in sexual behaviour in female guinea pig, *Anim. Behav.*, **7**, 142–149.

Grossman, S.P. (1975), The role of the hypothalamus in the regulation of food and water intake, *Psychol. Rev.*, **82**, 200–224.

Grubb, P. and Jewell, P.A. (1966), Social grouping and home range in feral Soay sheep. *Symp. of Zool. Soc. Lond.*, **18**, 179–210.

Grunt, J.A. and Young, W.C. (1952), Differential reactivity of individuals and the response of the male guinea pig to testosterone propionate, *Endocrinology*, **51**, 237–248.

Guhl, A.M. (1964), Psychophysiological interrelations in the social behaviour of chickens, *Psychol. Bull.*, **61**, 277–285.

Guhl, A.M., Collias, N.E. and Allee, W.C. (1945), Mating behaviour and the social hierarchy in small flocks of White Leghorns, *Physiol. Zool.*, **18**, 365–390.

Guhl, A.M. and Ortman, L.L. (1953), Visual patterns in the recognition of individuals among chickens, *Condor*, **55**, 287–298.

Guhl, A.M. and Warren, D.C. (1946), Number of offspring sired by cockerels related to social dominance in chickens, *Poult. Sci.*, **25**, 460–472.

Guiton, P. (1959), Socialization and imprinting in Brown Leghorn chicks, *Anim. Behav.*, **7**, 26–34.

Guiton, P. (1961), The influence of imprinting on the agonistic and courtship responses of the Brown Leghorn cock, *Anim. Behav.*, **9**, 167–177.

Gustavson, C.R., Kelly, D.J., Sweeney, M. and Garcia, J. (1976), Preylithium aversions. I: Coyotes and wolves, *Behav. Biol.*, **17**, 61–72.

Guyomarc'h, J.-CH. (1962), Contribution a l'etude du comportement vocal du poussin de 'Gallus domesticus', *J. Psychol. Norm. Path.*, **3**, 283–306.

Hafez, E.S.E. and Bouissou, M.-F. (1975), The behaviour of cattle, in *The Behaviour of Domestic Animals*, 3rd edn (Hafez, E.S.E., ed.). Bailliere Tindall, London. pp. 208–245.

Hale, E.B. (1962), Domestication and the evolution of behaviour, in *The Behaviour of Domesticated Animals*, 1st edn (Hafez, E.S.E., ed.). Bailliere Tindall & Cox, London, pp. 21–53.

Hale, E.B. (1966), Visual stimuli and reproductive behaviour in bulls, *J. Anim. Sci.*, **25** (suppl.), 36–44.

Hale, E.B. (1969), Domestication and the evolution of behaviour of animals, in *The Behaviour of Domestic Animals*, 2nd edn (Hafez, E.S.E., ed). Bailliere Tindall and Cassell, London.

Hall, S.J.G. (1979), Studying the Chillingham wild cattle, *Ark*, **6**, 72–79.

Hamilton, J.B. (1938), Precocious masculine behaviour following administration of synthetic male hormone, *Endocrinology*, **23**, 53–57.

Harlow, H.F. (1959), Love in infant monkeys, *Sci. Am.*, **200**, 68–89.

Harper, A.E. (1976), Protein and amino acids in the regulation of food intake, in *Hunger: Basic Mechanisms and Clinical Implications* (Novin, D., Wyrwicka, W. and Bray, G.A., eds). Raven Press, New York, pp. 103–113.

Harris, G.W. and Michael, R.P. (1964), The activation of sexual behaviour by hypothalamic implants of oestrogen, *J. Physiol.*, **171**, 275–301.

Hemsworth, P.H. and Beilharz, R.G. (1979), The influence of restricted physical contact with pigs during rearing on the sexual behaviour of the male domestic pig, *Anim. Prod.*, **29**, 311–314.

Hetherington, A.W. and Ranson, J.W. (1940), Hypothalamic lesion and adiposity in the rat, *Anat. Rec.*, **78**, 149–172.

Hilgard, E.R. (1957), *Introduction to Psychology*, 2nd edn. Harcourt, Brace & Co., New York.

Hinde, R.A. (1956), The behaviour of certain Cardueline F_1 inter-species hybrids, *Behaviour*, **9**, 202–214.

Hinde, R.A. (1969), The bases of aggression in animals, *J. Psychosomat. Res.*, **13**, 213–219.

Hinde, R.A. (1970), *Animal Behaviour. A Synthesis of Ethology and Comparative Psychology*, 2nd edn. McGraw-Hill Kogakusha Ltd, London.

Houpt, K.A., Law, K. and Martinisi, V. (1978), Dominance hierarchies in domestic horses, *Appl. Anim. Ethol.*, **4**, 273–284.

Huffaker, C.B. and Smith, R.F. (1980), Rationale, organization and development of a National Integrated Pest Management project, in *New Technology of Pest Control* (Huffaker, C.B., ed.). John Wiley & Sons Chichester.

Hughes, B.O. (1970a), Unpublished PhD thesis, University of Edinburgh.

Hughes, B.O. (1970b), Allelomimetic feeding in the domestic fowl, *Brit. Poult. Sci.*, **12**, 359–366.

Hughes, B.O. (1975), Spatial preference in the domestic hen, *Brit. Vet. J.*, **131**, 560–564.

Hughes, B.O. (1976a), *Behaviour as an index of welfare*. Vth European Poult. Conf., pp. 1005–1014.

Hughes, B.O. (1976b), Preference decisions of domestic hens for wire or litter floors, *Appl. Anim. Ethol.*, **2**, 155 –165.

Hughes, B.O. (1977), Selection of group size by individual laying hens, *Brit. Poult. Sci.*, **18**, 9–18.

Hughes, B.O. and Black, A.J. (1973), The preference of domestic hens for different types of battery cage floor, *Brit. Poult. Sci.*, **14**, 615–619.

Hughes, B.O. and Duncan, I.J.H. (1972), The influence of strain and environmental factors upon feather pecking and cannibalism in fowls, *Brit. Poult. Sci.*, **13**, 525–547.

Hughes, B.O. and Wood-Gush, D.G.M. (1971), A specific appetite for calcium in domestic fowls, *Anim. Behav.*, **19**, 490–499.

Hughes, B.O. and Wood-Gush D.G.M. (1977), Agonistic behaviour in domestic hens: the influence of housing method and group size, *Anim. Behav.*, **25**, 1056–1062.

Hunter, R.F. (1964), Home range behavioural hill sheep, in *Grazing in Terrestrial and Marine Environments* (Crisp, D.J., ed.). Blackwells, Oxford, pp. 155–172.

Hunter, R.H.F. (1980), *Physiology and Technology of Reproduction in Female Domestic Animals*. Academic Press, London.

Hunter, R.F. and Milner, C. (1963), The behaviour of individual, related and groups of South Country Cheviot hill sheep, *Anim. Behav.*, **11**, 507–513.

Hutton, R.C. (1979), *Social Facilitation as a Measure of Individual Relationships in a Pig Group (Sus scrofa L.)* Unpublished Honours thesis, Department of Zoology, University of Edinburgh.

Hutton, R.C., Stolba, A. and Wood-Gush D.G.M. (1980), Social facilitation as a measure of social relations in piglets kept extensively, *Appl. Anim. Ethol.*, **6**, 391.

Immelmann, K. (1975), The evolutionary significance of early experience, in *Function and Evolution in Behaviour*. Clarendon Press, Oxford.

Immelmann, K. (1980), *Introduction to Ethology*. Plenum Press, London.

Inglis, I.R. (1980), Visual bird scarers: an ethological approach, in *Bird Problems in Agriculture*. Proceedings of a Conference 'Understanding Bird Problems' (Wright, E.N., Inglis, I.R. and Feare, C.J., eds). BCPC

Publication, Croydon, pp. 121–143.

Jakway, J.S. (1959), The inheritance of patterns of mating behaviour in the male guinea pig, *Anim. Behav.*, 7, 150–162.

Johnson, J.I.Jr. (1975), States of activation: sleep, arousal and exploration, in *The Behaviour of Domestic Animals*, 3rd edn (Hafez, E.S.E., ed.). Bailliere Tindall, London, pp. 63–72.

Katongole, C.B., Naftown, F. and Short, R.V. (1974), Luteinizing hormone and testosterone levels in rams, *J. Endocr.*, 60, 101–106.

Kawamara, S. (1962), The process of sub-culture propagation among Japanese Macaques, *J. Primatol.*, 2, 43–60.

Keiper, R.R. (1970), Studies of stereotypy function in the canary (*Serinus canarius*), *Anim. Behav.*, 18, 353–357.

Kelley, D.B. and Pfaff, D.W. (1978), Generalizations from comparative studies on neuro-anatomical and endocrine mechanisms of sexual behaviour, in *Biological determinants of sexual behaviour* (Hutchison, J.B., ed.). John Wiley & Sons, Chichester, pp. 225–254.

Keverne, E.B. (1976), Sexual receptivity and attractiveness in the female rhesus monkey, in *Advances in Animal Behaviour*, 7. (Rosenblatt, J.S., Hinde, R.A., Shaw, E. and Beer, C., eds). Academic Press, London, pp. 155–200.

Kiley, M. (1972), The vocalizations of ungulates, their causation and function, 2, *Tierpsychol.*, 31, 171–222.

Kiley-Worthington, M. (1977), *Behavioural Problems of Farm Animals*. Oriel Press, London.

Kilgour, R. and De Langen, H. (1970), Stress in sheep resulting from management practices, *Proc. NZ Soc. Anim. Prod.*, 30, 65–76.

Kissileff, H.R. (1973), Non-homeostatic controls of drinking, in *The Neuropsychology of Thirst: New findings and Advances in Concepts* (Epstein, A.N., Kissileff, H.R. and Stellar, E., eds). Winston & Son, Washington DC, pp. 163–198.

Klingel, H. (1974), A comparison of the social behaviour of the Equidae, in *The Behaviour of Ungulates and Its Relation to Management* (Geist, V. and Walther, F., eds). Morges, Switzerland, Vol. 1, pp. 124–132.

Klingel, H. (1977), Observations on social organization and behaviour of African and Asiatic wild asses (*Equus africanus* and *E. hemionus*), *Z. Tierpsychol.*, 44, 323–331.

Klopfer, P.H. (1959), An analysis of learning in young Anatidae, *Ecology*, 40, 90–102.

Komai, T., Craig, J.V. and Wearden, J. (1959), Heritability and repeatability of social aggressiveness in the domestic chicken, *Poult. Sci.*, 38, 356–359.

Kratzer, D.D. and Craig, J.V. (1980), Mating behaviour of cockerels: effects of social status, group size and group density, *Appl. Anim. Ethol.*, 6, 49–62.

Kruijt J.P. (1964), Ontogeny of social behaviour in Burmese Red Jungle fowl (*Gallus gallus spadiceus*), *Behaviour* (Suppl.), **12**, 1–201.

Lawrence, A and Wood-Gush, D.G.M. (1982), Observations of lambs feeding on two different types of forage crops – rape and stubble turnips, *Appl. Anim. Ethol.*, **8**, 406–407.

Lea, R.W., Dods, A.S.M., Sharp, P.J. and Chadwick, A. (1981), The possible role of prolactin and the regulation of nesting behaviour and the secretion of luteinizing hormone in broody hens, *J. Endocr.*, **91**, 89–97.

Le Boeuf, B.J. and Petrinovich, L.F. (1974), Dialects of Northern Elephant Seals (*Mirounga angustirostris*): origins and reliability, *Anim. Behav.*, **22**, 656–663.

Lehrman, D.S. (1964), The reproductive behaviour of Ring Doves, *Sci. Am.*, **488**, 2–8.

Lent, P.C. (1974), Mother-infant relationships in ungulates, in *The Behaviour of Ungulates and Its Relation to Management* (Geist, V. and Walther, F., eds). IUCN, Morges, Switzerland, vol. 1, pp. 14–55.

Leung, P.M.B., Rodgers, Q.R. and Harper, A.E. (1968), Effect of amino acid imbalance on dietary choice in the rat, *J. Nutr.*, **95**, 483–492.

Leyhausen, P. and Heineman, I. (1975), 'Leadership' in a small herd of dairy cows, *Appl. Anim. Ethol.*, **1**, 206.

Lill, A. (1968), An analysis of sexual isolation in the domestic fowl. II: The basis of homogamy in females, *Behaviour*, **30**, 127–145.

Lill, A. and Wood-Gush, D.G.M. (1965), Potential ethological isolating mechanisms and assortative mating in the domestic fowl, *Behaviour*, **25**, 16–43.

Lindsay, D.R. and Fletcher, I.C. (1968), Sensory involvement on the mating behaviour of domestic sheep, *Anim. Behav.*, **16**, 410–414.

Lindsay, D.R. and Fletcher, I.C. (1972), Ram-seeking activity associated with oestrus behaviour in ewes, *Anim. Behav.*, **20**, 452–456.

Linkie, D.M. and Niswender, G.D. (1972), Serum levels of prolactin, luteinizing hormone and follicle stimulating hormone during pregnancy in the rat, *Endocrinology*, **90**, 632–637.

Lloyd Morgan, C. (1896), The habit of drinking in young chicks, *Science*, **3**, 900.

Lorenz, K. (1966), *On Aggression*. Methuen & Co. Ltd, London.

Lorenz, K. (1971), *Studies in Animal and Human Behaviour*, vol. 2. Methuen & Co. Ltd, London.

Lorenz, K (1981), *The Foundations of Ethology*. Springer Verlag, New York.

McBride, G. (1968), Social organization and stress in animal management, *Proc. Ecol. Soc. Aust.*, **3**, 133–138.

McBride, G., Parer, L.P. and Foenander, F. (1969), The social organization and behaviour of the feral domestic fowl, *Anim. Behav. Monogr.*, **2**, 127–181.

MacDonald, S.D. (1970), The breeding behaviour of the rock Ptarmigan,

Living Bird, **9**, 195–238.

McFarland, D.J. (1965), On the causal and functional significance of displacement activities, *Z. Tierpsychol.*, **23**, 217–235.

McFarland, D.J. (1970), Adjunctive behaviour in feeding and drinking situations, *Rev. Comp. Anim.*, **1**, 64–73.

McGill, T.E. (1978), Genetic factors influencing the action of hormones in sexual behaviour, in *Biological Determinants of Sexual Behaviour* (Hutchison, J.B., ed.). John Wiley & Son, Chichester, pp. 7–28.

Mackay, P.C. and Wood-Gush, D.G.M. (1980), The responsiveness of beef calves to novel stimuli: an interaction between exploration and fear, *Appl. Anim. Ethol.*, **6**, 383–384.

Maier, N.R.F. (1949), *Frustration. The Study of Behaviour Without a Goal.* McGraw-Hill, New York.

Maple, T (1975), Fundamentals of animal social behaviour., in *The Behaviour of Domestic Animals*, 3rd edn (Hafez, E.S.E., ed.). Bailliere Tindall, London, pp. 171–181.

Manning, A.W. (1979), *An Introduction to Animal Behaviour*, 3rd edn. Edward Arnold, London.

Marshall, J.F. (1976), Neurochemistry of central monoamine systems as related to food intake, in *Appetite and Food Intake: Report of the Dahlem Workshop on Appetite and Food Intake* (Silverstone, T., ed.) Life Sciences Research Report 2, Berlin, pp. 43–63.

Marx, J.L. (1973), Insect control (1): Use of pheromones, *Science*, **181**, 736–737.

Mason, J.W. (1975), Emotion as reflected in patterns of endocrine integration, in *Emotions: Their Parameters and Measurement.* (Levi, L., ed.). Raven Press, New York, pp. 141–181.

Matthews, R.W. and Matthews, J.R. (1978), *Insect Behaviour* John Wiley & Sons, Chichester.

Meddis, R. (1975), On the function of sleep, *Anim. Behav.*, **23**, 676–691.

Miller, N.E. (1957), Experiments on motivation. Studies combining psychological, physiological and pharmacological techniques, *Science*, **126**, 1271–1278.

Miller, N.E. and Banuazizi, A. (1968), Instrumental learning by curarized rats of a specific visceral response, intestinal or cardiac, *J. Comp. Physiol. Psychol.*, **65**, 1–7.

Millikan G.C. and Bowman, R.I. (1973), Observations on Galapagos tool-using finches in captivity, *The Living Bird*, **6**, 23–41.

Mills, A. and Wood-Gush, D.G.M. (1982), Genetic analysis of frustration response in the fowl, *Appl. Anim. Ethol.*, in press.

Moltz, H. (1975), Maternal behaviour: some hormonal, neural and chemical determinants, in *The Behaviour of Domestic Animals* (Hafez, E.S.E., ed.). Balliere Tindall, London, pp. 146–170.

Morgan, P.D., Boundy, C.A.P., Arnold, G.W. and Lindsay, D.R. (1975),

The roles played by the senses of the ewe in the location and recognition of lambs, *Appl. Anim. Ethol.*, **1**, 139–150.

Morris, D. (1964), The response of animals to a restricted environment, *Symp. Zool. Soc. Lond.*, **13**, 99–118.

Muller-Schwarze, D. (1974), Social function of various scent glands in certain ungulates and the problems encountered in experimental studies of scent communication, in *Behaviour of Ungulates and Its Relation to Management* (Geist, V. and Walther, F., eds). IUCN, Morges, Switzerland, pp. 107–113.

Murphy, L.B. (1969), *A study of some factors affecting choice of nest site in the domestic fowl*. Unpublished Honours thesis, The School of Agriculture, University of Edinburgh.

Murphy, L.B. and Wood Gush, D.G.M. (1978), The interpretation of the behaviour of domestic fowl in strange environments, *Biol. Behav.*, **3**, 39–61.

Nault, L.R. and Styer, W.E. (1972), Effects of sinigrin on host selection by aphids, *Entomol. Exp. Appl.*, **15**, 423–437.

Neuringer, A.J. (1970), Animals respond for food in the presence of free food, *Science*, **166**, 399–400.

Nissen, H.W., Chow, K.L. and Semmes, J. (1951), Effects of restricted opportunity for tactile, kinesthetic and manipulative experience on the behaviour of a chimpanzee, *Am. J. Psychol.*, **64**, 485–507.

Ookawa,T. and Gotoh, J. (1964), Electroencephalographic study of chickens: periodic recurrence of low voltage and fast waves during behavioural sleep, *Poult. Sci.*, **43**, 1603–1604.

Opel, H. and Proudman, J.A. (1980), Failure of mammalian prolactin to induce incubation behaviour in chickens and turkeys, *Poult. Sci.*, **59**, 2550–2558.

Oswald, I. (1980), Sleep as a restorative process: human clues, in *Adaptive Capabilities of the Nervous System. Progress in Brain Research*, Vol. 53 (McConnel, Boer Ronijn, van der Poll and Corner, eds). Elsevier, Amsterdam, pp. 279–288.

Perry, G. (1978), Behavioural needs, *Vet. Rec.*, **102**, 386–387.

Petney, T.N. and Bull, C.M. (1981), A non-specific aggregation pheromone in two Australian reptile ticks, *Anim. Behav.*, **29**, 181–185.

Pfaffman, C. and Bare, J.K. (1950), Gustatory nerve discharges in normal and adrenalectomized rats, *J. Comp. Physiol. Psychol.*, **43**, 320–324.

Phillips, R.E. and van Tienhouen, A. (1960), Endocrine factors involved in the failure of pintail ducks (*Anas acuta*) to reproduce in captivity, *J. Endocrin.*, **21**, 253–261.

Poindre, P. and Le Neindre, P. (1980), Endocrine and sensory regulation of maternal behaviour in the ewe, in *Advances in the Study of Behaviour. II* (Rosenblatt, J.S., Hinde, R.A., Beer, C. and Busnel, M.C., eds). Academic Press, London, pp. 76–120.

Powley, T.L. (1977), The ventromedial hypothalamic syndrome, satiety, and a cephalic phase hypothesis, *Psychol. Rev.*, **84**, 89–126.

Räber, H. (1948), Analyse des Balzverhaltens eines domestizieren Trutahns (*Meleagris*), *Behaviour*, **1**, 237–267.

Ramsay, A.O. (1953), Variations in the development of broodiness in fowl, *Behaviour*, **5**, 51–57.

Ramsay, A.O. (1961), Behaviour of some hybrids in the Mallard group, *Anim. Behav.*, **9**, 104–105.

Rattner, S.C. and Boice, R. (1975), Effects of domestication on behaviour, in *The Behaviour of Domestic Animals*, 3rd edn (Hafez, E.S.E., ed.). Bailliere Tindall, London, pp. 3–19.

Rheingold, H.L. and Hess, E.H. (1957), The chicks 'preferences' for some visual properties of water, *J. Comp. Physiol. Psychol.*, **50**, 417–421.

Richter, C.P. and Eckert, J.F. (1937), Increased calcium appetite of parathyroidectomized rats, *Endocrin.*, **21**, 50–54.

Riddle, O., Bates, R.W. and Lahr, E.L. (1935), Prolactin induces broodiness in fowl, *Am. J. Physiol.*, **111**, 352–360.

Richardson, R.H., Ellison, J.R. and Averhoff, W.W. (1982), Control of screw worm in North America, *Science*, **215**, 361–370.

Riehl, L.A., Brooks, R.F., McCoy, C.W., Fisher, T.W. and Dean, H.A. (1980), Accomplishments toward improving integrated pest management for citrus, in *New Technology of Pest Control*, (Huffaker, C.B., ed.). John Wiley & Sons, Chichester, pp. 319–364.

Roelofs, W.L. (1975), Insect communication – chemical, in *Insects, Science and Society* (Pimental, D., ed.). Academic Press, New York, pp. 79–99.

Rolls, E.T. (1976), Neurophysiology of feeding, in *Appetite and Food Intake: Report of the Dahlem Workshop on Appetite and Food Intake* (Silverstone, T., ed.). Life Sciences Research Report 2, Berlin, pp. 21–42.

Rompré, P.-P. and Miliaressis, E. (1980), A comparison of the excitability cycles of the hypothalamic fibres involved in self-stimulation and exploration, *Physiol. and Behav.*, **24**, 995–998.

Rosenblatt, J.S. and Aronson, L.R. (1958), The decline of sexual behaviour in male cats after castration with special reference to the role of prior sexual experience, *Behaviour*, **12**, 285–339.

Rosenblatt, J.S., Siegal, H.I. and Mayer, A.D. (1979), Progress in the study of maternal behaviour int he rat: hormonal, non-hormonal, sensory and developmental aspects, in *Advances in the Study of Behaviour*, **10** (Rosenblatt, J.S., Hinde, R.A., Beer, C. and Busnel, M.C., eds). Academic Press, London, pp. 226–311.

Rowland, N. (1980), Drinking behaviour: physiological, neurological and environmental factors, in *Analysis of Motivational Processes* (Toates, E.M. and Halliday, T.R., eds). Academic Press, London, pp. 39–59.

Rozin, P. (1967), Specific aversions as a component of specific hungers, *J. Comp. Physiol. Psychol.*, **64**, 237–242.

Rozin, P. (1976), The selection of foods by rats, humans and other animals, in *Advances in the Study of Behaviour*, 6 (Rosenblatt, J.S., Hinde, R.A., Shaw, E. and Beer, C., eds). Academic Press, London, pp. 21–76.

Ruckebusch, Y. (1972), The relevance of drowsiness in the circadian cycle of farm animals, *Anim. Behav.*, 20, 637–643.

Sachs, B.D. and Barfield, R.J. (1976), Functional analysis of masculine copulatory behaviour in the rat, in *Advances in the Study of Animal Behaviour*, 7 (Rosenblatt, J.S., Hinde, R.A., Shaw, E. and Beer, C., eds). Academic Press, London, pp. 92–154.

Salomon,A.L., Lazorcheck, M.J. and Schein, M.W. (1966), Effect of social dominance on individual crowing rates of cockerels, *J. Comp. Physiol. Psychol.*, 61, 144–146.

Sambraus, H.H. (1977), Observations and experiments on social behaviour on a herd of cattle during an 11-year period, *Appl. Anim. Ethol.*, 3, 199–200.

Sambraus, H.H. and Sambraus, D. (1975), Prägung von Nutztieren auf Menschen, *Z. Tierpsychol.*, 38, 1–17.

Sauer, C.O. (1952), *Agricultural Origins and Dispersals*. American Geographic Society, New York.

Schäfer,M. (1975), *The Language of the Horse: Habits and Forms of Expression*. Kaye & Ward, London.

Schein, M.W. and Fohrman, M.H. (1955), Social dominance in a herd of dairy cattle, *Br. J. Anim. Behav.*, 3, 45–55.

Schloeth, R. (1956),Quelques moyens d'intercommunication des taureaux de Camargue, *La Terre et la Vie*, 103, 83–93.

Schloeth, R. (1958), Cycle annuel et comportement social du taureau de Camargue, *Mammalia*, 22, 121–139.

Scott, E.M. and Quint, E. (1946), Self-selection of diet. III: Appetites for B vitamins, *J. Nutr.*, 32, 285–292.

Scott, J.P. (1950), The social behaviour of dogs and wolves: an illustration of sociobiological systematics, *Ann. NY Acad. Sci.*, 51, 1009–1021.

Scott, J.P. (1962), Critical periods in behavioural development, *Science*, 138, 949–958.

Selman, I.E., McEwan, A.D. and Fisher, E.W. (1970), Studies on natural suckling in cattle during the first eight hours post-partum. II: Behavioural studies (calves), *Anim. Behav.*, 18, 282–289.

Selye, H. (1960), The concept of stress in experimental physiology, in *Stress and Psychiatric Disorders* (Tanner, J.M., ed.). Blackwell Scientific Publications, Oxford. pp. 67–75.

Seyfarth, R.M. (1980), The distribution of grooming and related behaviours among adult female Vervet monkeys, *Anim. Behav.*, 28, 798–813.

Sharpe, R.S. and Johnsgard, P.A. (1966), Inheritance of behavioural characters on F_2 Mallard × Pintail (*Anas playrhynches* L. × *Anas aclita* L.) hybrids, *Behaviour*, 27, 259–272.

Shemesh, M., Ayalon, N. and Linder, H.R. (1972), Oestradiol levels in the peripheral blood of cows during the oestrus cycle, *J. Endocr.*, 55, 73–78.

Sheppard, P.M. (1951), Fluctuations in the selective value of certain phenotypes in the polymorphic land snail (*Cepea nemoralis* L.) *Heredity*, 5, 125–134.

Shillito, E.E. and Alexander, G. (1975), Mutual recognition amongst ewes and lambs of four breeds of sheep (*Ovis aries*), *Appl. Anim. Ethol.*, 1, 151–165.

Siegal, P.B. (1972), Genetic analysis of male mating behaviour in chickens (*Gallus domesticus*). I: Artificial selection, *Anim. Behav.*, 20, 564–570.

Signoret, J.-P. (1971), The reproductive behaviour of pigs in relation to fertility, *Vet. Rec.*, 88, 34–38.

Slater, P.J.B. (1980), Bird behaviour and scaring by sounds, *Bird Problems in Agriculture*. Proceedings of a Conference 'Understanding Bird Problems' (Wright, E.N., Inglis, I.R. and Feare, C.J., eds). BCPC Publication, Croydon, pp. 105–114.

Sly, J. and Bell, F.R. (1979), Experimental analysis of the seeking behaviour observed in ruminants when they are sodium deficient, *Physiol. Behav.*, 22, 499–505.

Smith, F.V., Van Toller, C. and Boyes, T. (1966), The 'critical period' in the attachment of lambs and ewes, *Anim. Behav.*, 14, 120–125.

Smith, W. (1957), Social 'learning' in domestic chicks, *Behaviour*, 11, 40–55.

Smith, W. and Hale, E.B. (1959), Modification of social rank in the domestic chicken, *J. Comp. Physiol. Psychol.*, 52, 373–375.

Smythies, J.R. (1970), *Brain Mechanisms and Behaviour*, 2nd edn. Blackwells Scientific Publications, Oxford.

Sonneman, P. and Sjolander, S. (1977), Effects of cross-fostering on the sexual imprinting of the female finch *Taeniopygia guttata*, *Z. Tierpsychol.*, 45, 337–348.

Spalding, D. (1873), 'Instinct' Macmillan's Magazine, 27, 282–93. Reprinted 1954, *Br. J Anim. Behav.*, 2, 2–11.

Spanier, E. (1980), The use of distress calls to repel night herons (*Nycticorax nycticorax*) from fish ponds, *J. Appl. Ecol.*, 17, 287–294.

Stadie, C. (1968), Verhaltensweisen von Gattungsbastarden *Phasianus corchicus* × *Gallus gallus* f. domestica in Vergleich mit denen der Ausgangsarten, *Verh. Dtsch. Zool. Ges*, (1967) 493–510.

Stellar, E. (1954), The physiology of motivation, *Psychol. Rev.*, 61, 5–22.

Stephens, D.B. (1975), Effects of gastric loading on the sucking response and voluntary milk intake in neonatal piglets, *J. Comp. Physiol. Psychol.*, 88, 796–805.

Stolba, A. (1982), A family system of pig housing. *Alternatives to Intensive Husbandry Systems*. UFAW Symp., London.

Stolba, A. and Wood-Gush, D.G.M. (1980), Arousal and exploration in growing pigs in different environments, *Appl. Anim. Ethol.*, 6, 382–383.

226   *Elements of Ethology*

Stolba, A. and Wood-Gush, D.G.M. (1981), The assessment of behavioural needs of pigs under free-range and confined conditions, *Appl. Anim. Ethol.*, 8, 583.

Stolba, A and Wood Gush, D.G.M. (1982), Verhaltungsgliederung und Reaktion auf 'Neureize als ethologische Kriterion zur Beurteilung von Haltungs bedingungen bei Hausschweinen, in *Aktuelle Arbeiten zur artgemassen Tierhaltung* (1980), 264, KTBL-Schrift.

Stricker, E.M. (1973), Thirst, sodium appetite and complementary physiological contributions to the regulation of intravascular fluid volume, in *The Neuropsychology of Thirst: New Findings and Advances in Concepts* (Epstein, A.N., Kissileff, H.R. and Stellar, E., eds). Winston & Sons, Washington DC. pp. 73–98.

Strubbe, J.H. and Mein, C.G. (1977), Increased feeding in response to bilateral injection of insulin antibodies in the V.M.H., *Physiol. Behav.*, 19, 309–313.

Strubbe, J.H. Steffens, A.B. and De Ruiter, L. (1977), Plasma insulin and the time pattern of feeding in the rat, *Physiol. Behav.*, 18, 81–86.

Syme, G.J. and Syme, L.A. (1975), The concept of spatial leadership in farm animals: an experiment with sheep, *Anim. Behav.*, 23, 921–923.

Ter Pelkwijk, J.J. and Tinbergen, N. (1937), Eine reizbiologische Analyse einiger Verhaltungsweisen von *Gasterosteus aculeatus* L, *Z. Tierpsychol.*, 1, 193–204.

Thorpe, W.H. (1958), Further studies on the process of song learning in the chaffinch (*Fringilla coelebs gergleri*), *Nature, Lond.*, 182, 554–557.

Thorpe, W.H. (1961), *Bird-song. The Biology of Vocal Communication and Expression in Birds*, Cambridge University Press, Cambridge..

Thorpe, W.H. (1963), *Learning and Instinct in Animals*, 2nd edn. Methuen & Co. Ltd, London.

Thorpe, W.H. (1965), The assessment of welfare of animals in intensive livestock husbandry systems, in *The Report of the Technical Committee to Enquire into the Welfare of Animals Kept under Intensive Livestock Systems*, Command Paper 2896. HMSO, London, pp. 71–79.

Tinbergen, N. (1951), *The Study of Instinct*. Oxford University Press, London.

Tinbergen, N. (1952), 'Derived Activities': their causation, biological significance, origin and emancipation during evolution. *Quart. Rev. Biol.*, 27, 1–132.

Tinbergen, N. (1959), Comparative studies of the behaviour of gulls (Laridae): a progress report, *Behaviour*, 15, 1–70.

Tinbergen, N. and Perdeck, A.C. (1950), On the stimulus situation releasing the begging response in the newly-hatched herring gull chick (*Larus a. argentatus* Pont.), *Behaviour*, 3, 1–38.

Tindell, D. and Craig, J.V. (1959), Effects of social competition on laying-house performance in the chicken, *Poult. Sci.*, 38, 95–105.

Toates, F.M. (1979), Homeostatis and drinking, *Behav. Brain Sci.*, **2**, 95–139.

Tolman, C.W. and Wilson, G.F. (1965), Social feeding in domestic chicks, *Anim. Behav.*, **13**, 134–142.

Trevorrow, R.L., Stone, B.F. and Cowio, M. (1977), Aggregation pheromones in two Australian hard ticks (*Ixodes holocyclus* and *Aponomma concolor*), *Experientia*, **33**, 680–682.

Tribe, D. (1950), The compositon of a sheep's natural diet, *J. Br. Grassld Soc.*, **5**, 81–92.

Tschanz, B. and Hirsbrunner-Scharf, M. (1975), Adaptations to colony life on cliff ledges: a comparative study of guillemot and razorbill chicks, in *Function and Evolution in Behaviour* (Baerends, G., Beer, C. and Manning, A., eds). Clarendon Press, Oxford, pp. 358–380.

Tyler, S.J. (1972), The behaviour and social organisation of the New Forest ponies, *Anim. Behav. Monogr.*, **5**, 85–196.

Valenta, J.G. and Rigby, M.K. (1968), Discrimination of the odour of stressed rats, *Science*, **161**, 599–601.

Van Iersel, J.J. and Bol, A.C.A. (1958), Preening in two tern species. A study of displacement activities, *Behaviour*, **13**, 1–88.

Van Rhijn, J.G. (1977), The patterning of preening and other comfort behaviour in a herring gull, *Behaviour*, **63**, 71–109.

Vestergaard, K. (1978), Normal behaviour of egg-laying birds. *1st Danish Seminar on Poultry Welfare in Egg-laying Cages*. National Committee for Poultry and Eggs, Copenhagen, pp. 11–17.

Vidal, J.M. (1980), The relations betwen filial and sexual imprinting in the domestic fowl: effects of age and social experience, *Anim. Behav.*, **28**, 000 001.

Vince, M.A. (1964), Synchronization of hatching in the Bobwhite quail (*Colinus virginianus*), *Nature, Lond.*, **208**, 1192–1193.

Vincent, L.E. and Bekoff, M. (1978), Quantitative analysis of the ontongeny of predatory behaviour in coyotes (*Canis latrans*), *Anim. Behav.*, **26**, 225–231.

Von Üexkuëll, J. (1957), A stroll through the world of animals and man, in *Instinctive Behaviour* (Schiller, C.H., ed.). International University Press, New York, pp. 5–80.

Walther F. (1974), Some references on expressive behaviour in combat and courtship of certain horned ungulates, in *Behaviour of Ungulates and Its Relation to Management* I (Geist, V. and Walther, F., eds). IUCN, Morges, Switzerland, pp. 56–106.

Waring G.H., Wierzbowski, S. and Hafez, E.S.E. (1975), The behaviour of horses, in *The Behaviour of Domestic Animals*, 3rd edn (Hafez, E.S.E., ed.). Bailliere Tindall, London, pp. 330–369.

Watson, A. (1972), The behaviour of Ptarmigan, *Brit. Birds*, **65**, 6–26; 93–117.

Weingarten, H. and White, N. (1978), Exploration evoked by electrical stimulation of the amygdala of rats, *Physiol. Psychol.*, **6**, 229–235.

Weir, W.C. and Torrell, D.C. (1959), Selective grazing by sheep as shown by a comparison of the chemical composition of range and pasture forage obtained by hand clipping and that collected by oesophageal-fistulated sheep, *J. Anim. Sci.*, **18**, 641–649.

Wiepkema, P.R. (1971), Positive feedbacks at work during feeding, *Behaviour*, **39**, 266–273.

Wilson, E.O. (1975), *Sociobiology. The New Synthesis*. The Belknap Press of Harvard University Press, London.

Wood-Gush, D.G.M. (1954), Observations on the nesting habits of Brown Leghorn hens, Section Paper *10th World Poultry Congress*, Department of Agriculture, Edinburgh, Scotland, pp. 187–192.

Wood-Gush, D.G.M. (1956), The agonistic and courtship behaviour of the Brown Leghorn cock, *Br. J. Anim. Behav.*, **4**, 133–142.

Wood-Gush, D.G.M. (1958a), Fecundity and sexual receptivity in the Brown Leghorn hen, *Poult. Sci.*, **37**, 30–33.

Wood-Gush, D.G.M. (1958b), The effect of experience on the mating behaviour of the domestic cock, *Anim. Behav.*, **6**, 68–71.

Wood-Gush, D.G.M. (1959), Time-lapse photography: a technique for studying diurnal rhythms, *Physiol. Zool.*, **32**, 272–283.

Wood-Gush, D.G.M. (1960), A study of the sex drive of two strains of cockerels through three generations, *Anim. Behav.*, **8**, 43–53.

Wood-Gush, D.G.M. (1963), The relationship between hormonally induced sexual behaviour in male chicks and their adult sexual behaviour, *Anim. Behav.*, **11**, 400–402.

Wood-Gush, D.G.M. (1971), *The Behaviour of the Domestic Fowl*. Heinemann Educational Books Ltd, London.

Wood-Gush, D.G.M. (1972) Strain differences in response to sub-optimal stimuli in the fowl, *Anim. Behav.*, **20**, 72–76.

Wood-Gush, D.G.M. (1973), Animal welfare in modern agriculture, *Br. Vet. J.*, **129**, 167–174.

Wood-Gush, D.G.M. (1975a), Nest construction by the domestic hen: some comparative and physiological considerations, *Neural and Endocrine Aspects of Behaviour in Birds* (Wright, P., Caryl, P.G. and Vowles, D.M., eds). Elsevier, Oxford, pp.35–49.

Wood-Gush, D.G.M. (1975b), The effect of cage floor modification on pre-laying behaviour, *Appl. Anim. Ethol.*, **1**, 113–118.

Wood-Gush, D.G.M. and Beilharz, R.G. (1983), The enrichment of a bare environment for animals in confined conditions, *Appl. Anim. Ethol.*, in press.

Wood-Gush D.G.M., Duncan, I.J.H. and Fraser, D. (1975), Social stress and welfare problems in agricultural animals, in *The Behaviour of*

Domestic Animals, 3rd edn (Hafez, E.S.E., ed.). Bailliere Tindall, London, pp. 182–200.

Wood-Gush, D.G.M., Duncan, I.J.H. and Savory, C.J. (1978), Observations on the social behaviour of domestic fowl in the wild, *Biol. Behav.*, 3, 193–205.

Wood-Gush, D.G.M. and Gilbert, A.B. (1964), The control of the nesting behaviour of the domestic hen. II: The role of the ovary, *Anim. Behav.*, 12, 451–453.

Wood-Gush, D.G.M. and Gilbert, A.B. (1969a), Oestrogen and the prelaying behaviour of the domestic hen, *Anim. Behav.*, 17, 586–589.

Wood-Gush, D.G.M. and Gilbert, A.B. (1969b), Observations on the laying behaviour of hens in battery cages. *Br. Poult. Sci.*, 10, 29–36.

Wood-Gush D.G.M. and Kare, M.R. (1966), The behaviour of calcium deficient chickens, *Br. J. Poult. Sci.*, 7, 285–290.

Wood-Gush, D.G.M. and Murphy, L.B. (1970), Some factors affecting the choice of nests by the hen, *Br. Poult. Sci.*, 11, 415–417.

Wood-Gush, D.G.M., Stolba, A. and Miller, C. (1982), Exploration in farm animals and animal husbandry, in *Exploration in Animals and Man* (Archer, J. and Birke, L., eds). Van Nostrand Reinhold, London.

Yerkes, R.M. and Morgulis, S. (1909), The method of Pavlov in animal psychology. *Psychol. Bull.*, 6, 257–273.

Zeuner, F.E. (1963), *A History of Domesticated Animals*. Harper & Row, New York.

SUBJECT INDEX

AUTHOR INDEX